Perceived Quality of Mobile Services

Forschungsergebnisse der Wirtschaftsuniversität Wien

Band 18

PETER LANG
Frankfurt am Main · Berlin · Bern · Bruxelles · New York · Oxford · Wien

Astrid Dickinger

Perceived Quality of Mobile Services

A Segment-Specific Analysis

PETER LANG
Internationaler Verlag der Wissenschaften

Bibliographic Information published by the Deutsche Nationalbibliothek
The Deutsche Nationalbibliothek lists this publication in the Deutsche Nationalbibliografie; detailed bibliographic data is available in the internet at <http://www.d-nb.de>.

Cover design:
Atelier Platen according to a design of the
advertising agency Publique.

University logo of the Vienna University of Economics
and Business Administration.
Printed with kind permission of the University.

Sponsored by the Vienna University of Economics
and Business Administration.

ISSN 1613-3056
ISBN 978-3-631-56487-5
© Peter Lang GmbH
Internationaler Verlag der Wissenschaften
Frankfurt am Main 2007
All rights reserved.

All parts of this publication are protected by copyright. Any utilisation outside the strict limits of the copyright law, without the permission of the publisher, is forbidden and liable to prosecution. This applies in particular to reproductions, translations, microfilming, and storage and processing in electronic retrieval systems.

www.peterlang.de

TABLE OF CONTENT

1 INTRODUCTION 1
1.1 Research Questions and Aims 3
1.2 Implications 5
1.3 Research Design 6
1.4 Structure 7

2 THE EXPLORATIVE RESEARCH 9
2.1 A Brief Summary of Qualitative Research 9
2.1.1 Advantages and Disadvantages of Qualitative Research Methods 11
2.1.2 Preparing for the Field and the Selection of Experts 12
2.1.3 Qualitative Interviewing 13
2.1.4 Qualitative Data Processing and Analyzing 14
2.2 Carrying out the Qualitative Research 15

3 THE M-COMMERCE VALUE CHAIN 19
3.1 The Basic Model 19
3.2 Infrastructure and Services 20
3.2.1 Mobile Transport 20
3.2.2 Mobile Interface and Applications 21
3.2.2.1 Mobile Phones 22
3.2.2.1.1 Phones as Portable Entertainment Players 22
3.2.2.1.2 Phones as a New Marketing Tool for Retailers and Manufacturers ... 23
3.2.2.1.3 Phones as a Multi-Channel Shopping Device 23
3.2.2.1.4 Phones as Portable Navigation Guides 23
3.2.2.1.5 Phones as Tickets and Money 24
3.2.2.1.6 Phones as a Mobile Internet Device 24
3.2.3 Mobile Service Technologies and Delivery Support 24
3.2.3.1 Messaging Services 26
3.2.3.2 Security and Privacy 28
3.3 Content 28
3.3.1 Content Creation 28
3.3.2 Content Packaging 29
3.3.3 Market Making 30
3.4 The Mobile User 31

4 MOBILE SERVICES 35
4.1 Characteristics of Mobile Services 35
4.2 Types of Mobile Services 37
4.2.1 Information Services 37
4.2.2 Entertainment Services 38
4.2.3 Transaction Services 40
4.2.4 Communication Services 41
4.2.5 Mobile Marketing Services 41
4.2.5.1 Mobile Branding 42

	4.2.5.2	Mobile CRM 43
	4.2.5.3	Mobile Advertising 43
	4.2.5.4	Mobile Market Research 45

5 THEORETICAL AND METHODOLOGICAL FRAMEWORK 47

5.1 RELEVANCE OF THE THEORETICAL FRAMEWORK 47
5.2 DIFFUSION AND ADOPTION THEORIES 49
5.2.1 Diffusion of Innovations 50
5.2.2 Social Cognitive Theory 52
5.2.3 Theory of Planned Behavior & Theory of Reasoned Action 52
5.2.4 The Technology Acceptance Model 53
5.2.5 Task Technology Fit Model 54
5.2.6 Motivational Model 55
5.2.7 Unified Theory of Acceptance and Use of Technology 55
5.2.8 Comparative Analysis of Adoption Models 57
5.3 SERVICE QUALITY 58
5.3.1 Donabedian's Model 59
5.3.2 Grönroos' Model 60
5.3.3 Parasuraman, Zeithaml, Berry 61
5.3.4 Measurement of Service Quality 63
5.3.4.1 SERVQUAL 63
5.3.4.2 SERVPERF 64
5.3.5 Internet Quality 65
5.3.5.1 E-S-Quality 65
5.3.5.2 EC-SERVQUAL 67
5.3.5.3 WebQual by Barnes and Vidgen 67
5.3.5.4 WebQual by Loiacono, Watson and Goodhue 68
5.3.5.5 DeLone and McLean's Model of IS Success 68
5.3.5.6 Sitequal 69
5.3.5.7 eTailQ 69
5.3.5.8 Attitude Toward the Site – A_{ST} 70
5.4 BEHAVIORAL CONSEQUENCES OF PERCEIVED SERVICE QUALITY 71
5.4.1 Loyalty 71
5.4.1.1 Day's Two-Dimensional Loyalty Concept 71
5.4.1.2 Loyalty According to Jacoby, Chestnut and Kyrner 72
5.4.1.3 The Dick and Basu Loyalty Approach 72
5.4.1.4 Oliver's Dynamic Loyalty Perspective 74
5.4.1.5 Measuring Loyalty 76
5.4.1.6 The Relationship between Customer Satisfaction and Loyalty 77
5.4.2 Value 77
5.4.2.1 Utilitarian Approach to Perceived Value 78
5.4.2.2 Behavioral Approach to Perceived Value 79
5.5 DISCUSSION OF THE RELEVANT MODELS AND CONSTRUCTS 82
5.6 CAUSAL MODELING 88
5.6.1 Causal Models 89
5.6.2 Usage Areas for Causal Models 91
5.6.3 Steps for Structural Equation Modeling 92
5.6.3.1 Model Specification 92

	5.6.3.2 Model Identification	93
	5.6.3.3 Parameter Estimates	93
	5.6.3.4 Model Evaluation Measures	94
5.6.4	M-Plus	96
	5.6.4.1 Particularities Estimating Models with Categorical Data	96
	5.6.4.2 Estimators in Mplus	96
	5.6.4.3 Fit Indices for Categorical Variables	97

6 HYPOTHESES DEVELOPMENT AND MODEL SPECIFICATION 99

- 6.1 PERCEPTUAL ATTRIBUTES 101
- 6.2 HIGHER ORDER ABSTRACTIONS 103
- 6.3 BEHAVIORAL OUTCOMES 106
- 6.4 MODERATOR EFFECTS 107
  - 6.4.1 Innovativeness 107
  - 6.4.2 Experience 108
  - 6.4.3 Age 108

7 OPERATIONALIZATION OF THE CONSTRUCTS 109

8 DATA COLLECTION 113

- 8.1 QUESTIONNAIRE DESIGN 113
- 8.2 SAMPLING AND SAMPLE SIZE 114
- 8.3 PRETEST 115
- 8.4 THE FIELD PHASE 116

9 EMPIRICAL TEST AND ANALYSES 119

- 9.1 RESPONSE RATE 119
- 9.2 SAMPLE STRUCTURE 120
  - 9.2.1 Profile of m-parking Users and Non-Users 121
  - 9.2.2 M-Services Usage Behavior 122
  - 9.2.3 M-Parking Usage 125
  - 9.2.4 Data Structure 126
- 9.3 TEST RESULTS 127
  - 9.3.1 Test of the Hypothesized Model 127
    - 9.3.1.1 Test of the Measurement Model 130
  - 9.3.2 Using Dichotomous Variables for Model Test 134
    - 9.3.2.1 The Measurement Model Using Dichotomous Variables 137
  - 9.3.3 Summary and Interpretation of Findings from the Model Test 139
- 9.4 ALTERNATIVE MODELS 141
  - 9.4.1 Alternative 1 141
  - 9.4.2 Alternative 2 142
- 9.5 SEARCHING FOR HETEROGENEITY 143
  - 9.5.1 Multiple Group Analysis 144
    - 9.5.1.1 Innovativeness as Grouping Variable 145
    - 9.5.1.2 Experience as Grouping Variable 146
    - 9.5.1.3 Age as Grouping Variable 148
  - 9.5.2 Latent Class Analysis 149
    - 9.5.2.1 Model Differences 153

	9.5.2.2	Distinct Characteristics of the Classes Identified	154
		9.5.2.2.1 Class One: Mobile Service Skeptics	154
		9.5.2.2.2 Class Two: Undecided Users	155
		9.5.2.2.3 Class Three: Cautious Innovators	155
		9.5.2.2.4 Class Four: Mobile Service Lovers	156

10 CONCLUSIONS 157

  10.1 IMPLICATIONS FOR FUTURE RESEARCH 159
  10.2 IMPLICATIONS FOR INDUSTRY 163

11 REFERENCES 166

12 APPENDIX - QUESTIONNAIRE 187

LIST OF FIGURES

Figure 1: Differences Between Qualitative and Quantitative Methods 10
Figure 2: The m-commerce Value Chain 20
Figure 3: User Interface of the Chordiant 5 Mobile Marketing Director 30
Figure 4: Categorization of Wireless Advertising with Examples 44
Figure 5: Compeau and Higgins SCT Model 52
Figure 6: The Theory of Planned Behavior 53
Figure 7: The Technology Acceptance Model 54
Figure 8: Task Technology Fit Model 55
Figure 9: UTAUT Model 57
Figure 10: Grönroos' Service Quality Model 60
Figure 11: Service Quality Model of Parasuraman/Zeithaml/Berry 62
Figure 12: DeLone and McLean's Model of IS Success 68
Figure 13: eTailQ 70
Figure 14: Attitude Toward the Site 70
Figure 15: A Framework of Customer Loyalty 73
Figure 16: Types of Loyalty according to Dick and Basu 73
Figure 17: Effects of Price Comparison on Perceptions of Value 79
Figure 18: Means-End Model Relating Price, Quality, and Value 79
Figure 19: Consumer Value Hierarchy Model 80
Figure 20: Holbrook's Typology of Consumer Value 81
Figure 21: Primary Symbols Used in Path Analysis 89
Figure 22: Causal Model 90
Figure 23: A Structural Model for Mobile Services User Behavior 100
Figure 24: Link Provided on the Official m-parking Web Site 116
Figure 25: Link Provided on wien.at 117
Figure 26: Frequencies for Filling out the Questionnaire 119
Figure 27: Services m-parking Users Would Like to Use in the Future 123
Figure 28: Services m-parking Non-Users Would Like to Use in the Future ... 123
Figure 29: Reasons to Register for M-parking 126
Figure 30: Full Research Model 127
Figure 31: Thresholds of the Categorical Variables 131
Figure 32: Full Research Model With Dichotomous Variables 135
Figure 33: Thresholds of the Dichotomous Variables 137
Figure 34: Alternative Model for Mobile Services Usage 141
Figure 35: The Behavioral Model 142
Figure 36: Means of the Four Classes for Two Trust Measures 150

LIST OF TABLES

Table 1: Advantages and Disadvantages of Using Qualitative Research Methods 11
Table 2: Group of Interviewees and Their Background 16
Table 3: Key Mobile Network Technologies 21
Table 4: Key Mobile Service Technologies 25
Table 5: Taxonomy of Mobile Services Adapted from 36
Table 6: Models and Theories of Individual Acceptance 49
Table 7: Review of Model Comparisons 58
Table 8: Dimensions of Perceived Service Quality and Related Constructs 59
Table 9: Dimensions of e-SQ 66
Table 10: Loyalty Phases with Corresponding Vulnerabilities 75
Table 11: Measuring Loyalty According to Oliver 75
Table 12: Diffusion, Adoption and Service Quality Models in IS and M-Services Research .. 83
Table 13: Notation for Latent and Measurement Model 91
Table 14: Fit Indices 95
Table 15: Definition of the Constructs Included in the Research Model 101
Table 16: Operationalization of the Constructs 111
Table 17: Profile of the Whole Sample 120
Table 18: Characteristics of the Sub-Sample "Users" 121
Table 19: Profile of the Sub-Sample M-parking Non-Users 121
Table 20: Provider Distribution of the Respondents 122
Table 21: Mobile Services Already Used by the Respondents 122
Table 22: Cross Tabulation of Frequency of M-Service Use and Usage of M-parking 124
Table 23: Cross Tabulation of Duration of M-Service Use and Usage of M-parking 125
Table 24: Fit Indices for the Full Research Model 128
Table 25: Summary of the Findings of the Hypotheses Test 129
Table 26: Item Mean, Standard Deviation, Cronbach's Alpha, and Factor Loading 131
Table 27: Overview of the Reliability of the Scales 133
Table 28: Fit Indices for the Research Model with Dichotomous Measures 135
Table 29: Summary of the Findings of the Hypotheses Test 136
Table 30: Cronbach's Alpha, Factor Loading, Critical Ratio, and R^2 137
Table 31: Overview of the Reliability of the Scales 139
Table 32: Overview of the Reliability of the Scales for the Alternative Model 142
Table 33: Overview of the Reliability of the Scales for the Behavioral Model 143
Table 34: Chi-Square Change with Innovativeness as Grouping Variable 145
Table 35: Path Estimate Difference and Tolerance Interval for Innovativeness 146
Table 36: Chi-Square Change with Experience as Grouping Variable 147
Table 37: Path Estimate Difference and Tolerance Interval for Experience 147
Table 38: Chi-Square Change with Age as Grouping Variable 148
Table 39: Path Estimate Difference and Tolerance Interval for Age 148
Table 40: Deciding on the Number of Classes 150
Table 41: Class Counts and Proportions 151
Table 42: Latent Class Probability for Most Likely Latent Class Membership 152
Table 43: Path Estimates and Standard Errors for the Four Classes 153

1 INTRODUCTION

The following sections introduce this book and the underlying research problem. First the introduction gives an overview about the mobile communication industry, then, the research questions and aim of the book follow. The second sub-chapter presents the implications and the research design is illustrated. The introduction closes presenting the structure.

The introduction of mobile devices – mobile phones, Personal Digital Assistants (PDAs), handhelds, etc. – and shift from voice to data transfer has changed the telecommunication industry. The business potential is immense, with Siemens estimating that European earnings for mobile communication industries will triple by 2010 from 1999 (Hartmann and Büppelmann, 2001, 2). Furthermore, using mobile devices for commerce, m-commerce, extends existing e-commerce applications by overcoming limitations of time and place (Müller-Veerse, 1999, 20f). Based on the mentioned unique characteristics, Kleijnen and colleagues (Kleijnen, Ruyter et al., 2002) define m-commerce as:

"Any electronic transaction of information interaction conducted using a mobile device and mobile networks (wireless or switched public network) thereby guaranteeing customers virtual and physical mobility; leading to the transfer of real or perceived value in exchange for personalized, location-based information, services or goods".

There are other definitions for m-commerce, some focusing only on the wireless aspect. Sadeh (2002, 5) for instance defines m-commerce as *"the emerging set of applications and services people can access form their Internet-enabled mobile devices."*

Balasubramanian et al. (2002, 349) do not just introduce a definition but try to conceptualize the phenomenon of m-commerce. According to them *"it involves communication, either one-way or interactive between two or more humans, between a human (or humans) and one or more inanimate objects (such as databases), or between two of more inanimate objects (e.g., between devices).*

At least one of the parties engaged in the communication must be mobile, in the sense that his, her or its ability to communicate is not contingent on being at a fixed physical location at a particular point in time.

The ability to communicate must possess the potential to be continuously maintained for at least one of the parties during a substantial physical movement from one location to another.

The communication signals between parties must be primarily carried by electromagnetic waves, without direct sensory perception of the signals.

If humans are communicating, at least one seeks to benefit economically from the communication, either in the short or the long run. If the communication is entirely between inanimate objects, such communication must be ultimately aimed at creating economic benefits for a human or a firm."

For this book the understanding of m-commerce goes beyond acknowledging that a device is wireless, thus, the conceptualization of Balasubramanian et al. (2002) is a good starting point. The users, at the heart of this book, are now explored in more detail.

European end-users have a strong demand for mobile data services, are willing to use services like Multimedia Messaging Service (MMS), advanced location-based services or wireless offices with a mobile device (Hartmann and Büppelmann, 2001) and willing to spend additional money for these services. Consumers are predicted to increase expenses on their current mobile services by 69% and business users by 42% (Hartmann and Büppelmann, 2001, 3). As e-commerce has not reached the explosive growth figures predicted in the mid 1990s, the eyes of scholars and industry representatives are now set on the opportunities offered by wireless media. They envisage that the next phase of e-business growth will take place in mobile commerce (Varshney and Vetter, 2001). While some authors predict skyrocketing developments in m-commerce, others share a more cautious perspective, which is natural when a new technology emerges (Schuster, 2001).

Mobile data transfer remains low; voice usage still represents 90% of revenues for mobile net operators (NetValue, 2002, 4). Short Message Service (SMS) is the most popular form of data transfer with more than 10 billion messages sent world-wide each month (Cohn, 2001).

A comparison of slow adoption of WAP in Europe with the impressive adoption of i-mode services in Japan and the simple SMS based services in Scandinavia suggest that aggregate and technology based models are insufficient to explain the process of mobile services adoption and use (Pedersen, Leif et al., 2002). The adoption decisions of individual end-users must be better understood to predict and explain the use of mobile services.

There is a lack of adequate service quality delivered via mobile devices. If wireless channels are to be accepted by consumers, companies must shift the focus of m-business/m-commerce to m-service, all cues and encounters that occur before, during and after the transactions. To deliver superior services companies must understand how consumers perceive and evaluate their mobile services.

The aim of this research is to develop and test a model that explains mobile services usage behavior. The mobile medium offers various ways of communication between companies and end users and there are some critical factors to keep in eye for the successful use of this medium. The following chapters introduce to the research questions and theoretical implications, the practical implications, the research design and the structure of the book.

1.1 Research Questions and Aims

Still in an experimental phase, little research exists and businesses have little experience with mobile media offering direct communication with consumers, anytime and anyplace. One of the key characteristics is the two way information flow of mobile messages between sender and receiver. This interactivity suggests drawing upon theories in marketing, consumer behavior, psychology and diffusion to investigate business and personal adoption of mobile services (Balasubramanian, Peterson et al., 2002; Barwise, Elberse et al., 2002; Barwise and Strong, 2002; Damanpour, 1991; Davis, 1989; Hoffman and Novak, 1996; Jee and Lee, 2002; Loch, Straub et al., 2003; Newell and Lemon, 2001; Rodgers and Thorson, 2000; Zmud and Apple, 1992).

Consumer behavior regarding mobile services has not yet been subject of much research in Europe but industry analysts have high expectations regarding the willingness to adopt mobile services.

Balasubramanian et al. (2002) introduce a taxonomy of mobile marketing usage and raise consumer-oriented, marketer-oriented and public policy-oriented research questions. Barwise and Strong (2002) reflect on the right execution of mobile advertising and conclude highlighting a void in research in the field of acceptance of services.

Concepts and guidelines for developing mobile services are generally missing. The Wireless World Research Forum (WWRF) presented their *"Book of Visions"* on the future of wireless networks stating:

"It will become more and more important how the users perceive the service and the emotional impact and pleasure that the service creates and maintains." (WWRF, 2000)

There is little discussion on perceptions, emotions and pleasures created at the service and end-user level. Often vision papers elaborate technological requirements without discussion of the important end-user issues.

Thus, the central contribution of this research is the development of a model explaining customer perception of the service quality of mobile services and the resulting behavioral consequences. The analysis sheds light on the views and perceptions of customers receiving mobile services. This book describes the development, refinement, psychometric evaluation, properties, and potential applications of a mobile service assessment and behavior model. The aims of the research are to answer the following questions:

- What are the antecedents of mobile service quality?
- How do consumers differ in their perceptions and behavior with regard to mobile services?
- Can the proposed model explain mobile consumer behavior?
- What are the behavioral consequences of perceived quality in a mobile services setting?
- The research questions can be split in four part-aims for a more formal perspective. The aim can be content, methodology, practical, and scientifically oriented including answers of research questions in the following areas.

Research Aims Concerning the Content
- Analysis of the topic "Mobile Service Quality and Behavioral Outcomes"
- Which factors drive mobile user behavior in Austria?
- Are there differences in the motivators for adoption in different segments?
- To what extent do existing service quality and diffusion theories support theory development?

Methodological Research Aims
- Is the software program Mplus appropriate to estimate the proposed causal model?
- Does the data support the proposed model?
- Are a-priori grouping variables useful?
- Does the result of Latent Class Analysis reflect the results found through a-priori grouping?

Practical Aims
- What strategic implications derive for companies in order to offer good mobile service quality?
- What are the main antecedents of perceived service quality?
- What attributes do the services need to contain to be rated as high quality by specific target groups?
- What drives consumers' loyalty?
- Are there any particular segments that need to be targeted differently by companies?

1.2 Implications

This research project is relevant for players in the mobile business arena and will help solve practical problems. The main participants in mobile markets who will profit from this survey are network operators, content and services providers, technology vendors and appliance manufacturers (Lehmann and Lehner, 2001). Enterprises planning to offer mobile commerce services must be aware of the primary concerns of consumers. Such knowledge can help these companies increase the usage of mobile services (Hung, Ku et al., 2003) and learn about service quality evaluation. Additionally companies can understand the resistance to services adoption among users and benefit from this information in order to increase the services' quality evaluation based on the user needs identified in the empirical study. Once critical factors are known companies can develop better services and improve performance to fit customer needs.

Above that companies can develop effective marketing strategies to convince customers that mobile commerce is a convenient sales method and pricing strategies are found accordingly.

Mobile phone operators will gain information to better target their own clients. Operators are in a good position concerning mobile services, as they have information about their clients. It is important for the operators that the customers are satisfied otherwise they would need to handle a huge amount of complaints as they are the first point of contact for mobile phone users. Network operators integrate mobile data services in their range of products to increase average revenue per user. However, they still are not certain what particular customers seek in their services.

There are only few successful mobile content providers. Those have an interest in the research because it will contribute scientific answers to their every day problems. It is important for them to get suggestions on how to optimize the mobile content for their customers. They would receive empirically sound evidence on the quality of the service they provide.

With regard to theoretic implications the author creates a model for mobile service quality and consumer behavior. Due to the nature of the mobile medium, ubiquitous accessibility, existing models have to be reviewed and modified.

The body of knowledge from service quality[1], diffusion of innovations and acceptance[2] literature was reviewed and used as a basis for theory development. Guided by theory, modifications of the previously developed model will be presented. In addition, to test of the causal model multiple group analysis will give insights into different user segments.

The emploment of latent class analysis allows data driven segmentation and will help identify user groups with significant differences.

1.3 Research Design

The research design consists of two major research phases, qualitative and quantitative:

In the first step, players in the m-commerce value chain are interviewed. Through literature review and expert interviews successful mobile services, their quality and content are identified and evaluated. The antecedents of service quality in the mobile data service industry and the customers' behavior are analyzed. On basis of that a causal model for mobile service quality and behavioral consequences is identified.

The second and major research phase tests the causal model based on expert interviews and literature review. The quantitative research shows if the developed hypotheses can be corroborated or need to be abandoned. In addition differences between customer segments are identified. Demographic and psychographic criteria are included in the segmentation process.

It needs to be mentioned that the findings are of limited validity. They just refer to the sample explored and are only valid until new findings are explored. Popper gave up the demand for absolute cognition and held the view that there will not be a definite knowledge that is able to explain everything. He is taking away the pressure of perfection; according to Popper the most important aspect is continuous critical exploration (Popper, 1976, 103ff).

After testing of the stipulated hypotheses some may be abandoned. Thus, some factors first assumed to be of importance for the quality perception of the mobile service may be falsified. Also, grouping variables hypothesized to prove successful for arriving with significant group differences may have to be reconsidered. Further literature review allows theory guided model modification.

---

[1] Cronin, Brady et al., 2000; Cronin and Taylor, 1992; Cronin and Taylor, 1994; Donabedian, 2003; Grönroos, 1978; Grönroos, 2001; Kettinger and Lee, 1994; Loiacono, Chen et al., 2002; Oliva, Oliver et al., 1992; Oliver, 1997; Parasuraman, Zeithaml et al., 1988; Parasuraman, Zeithaml et al., 2005; Wolfinbarger and Gilly, 2003; Zeithaml, Berry et al., 1996; Zeithaml, Parasuraman et al., 2000.

[2] Aijzen, 1991; Ajzen, 2001; Ajzen and Fishbein, 1980; Battacherjee, 2000; Bayarmaa and Boalch, 1997; Compeau and Higgins, 1995; Compeau, Higgins et al., 1999; Davis, 1989; Davis, Bagozzi et al., 1992; Goodhue and Thompson, 1995; Koufaris, 2002; Legris, Ingham et al., 2003; Loiacono, Chen et al., 2002; Rogers, 1995; Venkatesh and Davis, 2000; Venkatesh, Morris et al., 2003.

For the explorative research, expert interviews and literature review are the underlying methods. In the course of the quantitative research, primary data is collected via an online survey. The causal model is tested using the software package Mplus. This software tool also includes multiple group functions and a latent class analysis procedure employed for the identification of groups within the sample.

1.4 Structure

The book consists of five major parts. Each of them will be explained in more detail in this section:

First, the introduction provides an overview of the research problem, the research design and the expected implications.

Second, the research approach of the explorative research phase is explained, followed by a discussion of the qualitative survey of this study. The next section presents the field of m-commerce, combined with findings of the qualitative survey. A general introduction to the topic of mobile services is given. This provides the reader with a state of the art introduction to mobile commerce, the mobile commerce value chain, and successful mobile services.

The following chapters provide the theoretical building blocks which have a bearing in service quality and consumer acceptance research. These building blocks are presented and discussed. Subsequently further constructs relevant for m-service user behavior are introduced. The constructs are drawn together to propose a conceptual model of mobile services usage. Based on this the full model and consequently the research hypotheses are developed. Apart from providing the theoretic basis for the model development also the methodological base is provided. Chapter 5.6 gives an introduction to causal modeling and its application in marketing research.

The chapters dealing with the empirical research phase provide the operationalization of the constructs, followed by a detailed description on how the quantitative research was carried out. They shed light on how the survey is conducted, and explain general research challenges, the sampling, and data collection process. The analysis first presents descriptive findings and insights into the sample profile. Then the hypothesized model is estimated. The measurement model is evaluated to find validity defects. A test of the model with dichotomized variables provides further insights in the quality of the measurement instrument. Through further literature review modified models are identified and also tested. Further analyses are carried out to detect heterogeneity. First a-priori criteria are chosen as grouping variables for multiple group analysis. Furthermore latent class analysis arrives with segments that are detected based on the data structure.

The concluding chapter discusses the results and summarizes the main findings. Furthermore, there is a discussion of implications for industry and future research. The book closes with a list of references and an appendix (questionnaire).

2 EXPLORATIVE RESEARCH

This chapter discusses qualitative research as it is relevant for the first research process, then the application of theory for this particular case follows.

2.1 A Brief Summary of Qualitative Research

Although there are vast differences between the qualitative and quantitative research there is no set of factors that allows distinguishing them as mutually exclusive. Before exploring one of them in more detail the difference of the two approaches are outlined. Figure 1 offers insights on the general characteristics and differences of each of them.

A main objective of qualitative research is to gain preliminary insights into decision problems and opportunities. It focuses on the collection of primary data from small samples of respondents by asking questions or observing. Qualitative research is used in exploratory designs. Quantitative research on the other hand places heavy emphasis on the use of formalized standard questions and predetermines response portions in questionnaires or surveys administered to a large number of respondents.

Qualitative researchers have to be well trained in interpersonal communication and have interpretive skills. They frequently use open-ended questions allowing for in-depth probing of the initial responses and specific observation techniques for analysis of behavior.

The non structured format of the questions and the small sample size limit the researcher's ability to generalize the qualitative data to larger segments of subjects. Nevertheless, qualitative data have important uses in understanding problems in areas of initial discovery and preliminary explanation. This kind of data can provide decision makers with initial ideas about specific problems or opportunities, theories and models or constructs (Hair, Busch et al., 2000, 216). In most explorative research attempts the raw data will be collected through qualitative data collection.

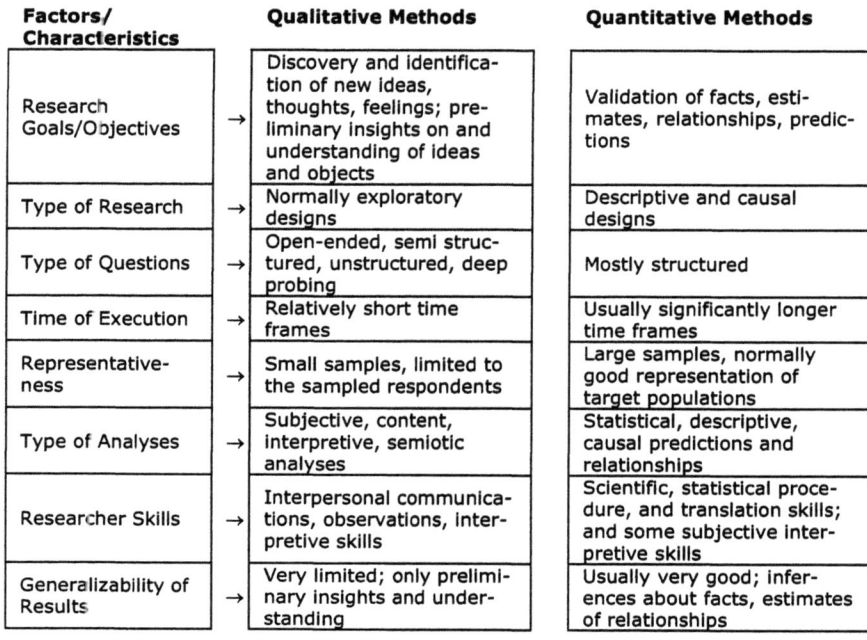

Figure 1: Differences Between Qualitative and Quantitative Research Methods (Hair, Busch et al., 2000)

Qualitative research methods for collecting and creating additional information are appropriate when researchers are (Hair, Busch et al., 2000, 217)

- In the process of identifying a business problem or opportunity or establishing information requirements
- Interested in obtaining some preliminary insights
- In the process of building theories and models to explain relationships, between two or more constructs
- Attempting to develop reliable and valid scale measurements for investigating specific market factors, consumer qualities and behavioral outcomes
- Interested in new-product or service development or repositioning current product or service images

In conclusion of above discussion on the main characteristics of qualitative research it can be summarized that: *„Qualitative research provides an in-depth insight; it is flexible, small-scale and exploratory and the results obtained are concrete, real-life like and full of ideas."* (Ruyter and Scholl, 1998)

2.1.1 Advantages and Disadvantages of Qualitative Research Methods

Qualitative research methods offer several advantages for today's researchers. One of the main advantages is the fact that qualitative research is economical and timely because of small samples. An other advantage is the rich and in-depth data about a subjects' attitudes, beliefs, emotions and perceptions (Hair, Busch et al., 2000). Such data can be invaluable to gaining a preliminary understanding of behaviors. Because of the richness of the qualitative data it can ideally supplement the facts and estimates gathered through other primary data collection methods. The key advantage is that the researcher does not have to rely on just reported behaviors but can accurately investigate and record actual behaviors.

Advantages of some qualitative research methods include that they provide insights into building marketing models, identify marketing problems and opportunities, enable researchers to predict consumer behavior, develop better marketing constructs, and more reliable and valid scale measurements.

Although useful information is gained through qualitative research, there are two major disadvantages. One is the sample size limitation and the other one is the need for well-trained interviewers, observers, and investigators.

Another disadvantage is the fact that small differences often the basis for marketing success and failure can not be detected by this sort of data.

Table 1: Advantages and Disadvantages of Using Qualitative Research Methods (Hair, Busch et al., 2000)

Advantages of Qualitative Methods	Disadvantages of Qualitative Methods
Economical and timely data collection Richness of the data Accuracy of recording marketplace behaviors	Lack of generalizability Inability to distinguish small differences Lack of reliability and validity
Preliminary insights into building models and scale measurements	Difficulty in finding well-trained investigators, interviewers and observers

In order to gain in depth insights into the phenomenon of interest it is essential to go into the field and observe the phenomenon where it is happening, participate to achieve detailed insights or interview people who are members of the group of interest. One form to gain data in qualitative research is the interview. In the following three subchapters the questions about what to obey when going into the field, how to do qualitative interviews, and how to analyze them will be discussed and explained.

2.1.2 Preparing for the Field and the Selection of Experts

If the researchers do not know too much about the group or phenomenon of interest they are well advised searching for relevant literature, filling in the knowledge of the subject, and learning about what others have said about it (Babbie, 1998, 289).

A next step is to make use of informants who have already studied the group of interest or with someone who is familiar with it. The relationship with the informant is a key indicator of the quality of the findings. Ideally the relationship goes beyond the researchers' research role.

The way the informants are approached and the role of the researcher influence the type of information gathered. Informants' knowledge is a mixture of facts and personal point of view (Babbie, 1998, 289).

Normally a researcher wants informants that are typical of the group studied. Otherwise their views and opinions might be misleading. However, researchers need to be cautious since simply because someone is willing to work with investigators makes him/her somewhat atypical within their group. The status of an informant might limit their access to the different sectors of the group under study as they might be *"marginal"* members.

In the underlying survey the view of experts - or to use a different terminology, the information of executives involved in the mobile industry - is of relevance. To get a coherent picture of the mobile commerce and mobile marketing industry various experts at all stages of the m-commerce value chain should be included in the interviewing process. This will involve interviews with executives holding positions in the industry of question. But what is an expert, or an executive that can be regarded as an expert? According to Rubin (1995) an expert is a person with in depth knowledge about the research topic. This knowledge can derive from working in that specific industry or by doing research in that field. The experts should be prepared to share their knowledge and talk about it. Above that the different experts' views should reflect the different aspects and opinions of the whole industry. Due to the fact that the knowledge of the individual expert is of critical relevance for the survey, the selection of the sample is crucial. Thus, it is appropriate to choose the sample on basis of your own knowledge about the group. This is called 'purposive sampling' (Shaw, 1999).

An executive interview often takes place in the surrounding the interviewee is used to and comfortable, which is his office or a conference room in the office building. According to Shaw (1999) it is important that the interview takes place in a surrounding the interviewee knows and is comfortable in. The interview should take place in a relaxed atmosphere to support the willingness of the interviewee to share his information (Kepper, 1996, 35).

Often it is very expensive to conduct executive interviews and getting an appointment can be time consuming (Hair, Busch et al., 2000, 257).

The number of interviews is depending on the complexity and the quality of data obtained in each interview. In case of 'theoretical saturation' – nothing new can be explored and there is no need for more interviews (Rubin and Rubin, 1995, 72). Due to the fact that in qualitative research not the number of equal statements is counted but the quality, richness, and content of the information provided is of relevance this procedure is feasible.

2.1.3 Qualitative Interviewing

As opposed to surveys with rigidly structured questionnaires, qualitative interview design is flexible, iterative and continuous. This indicates that the questionnaire is redesigned throughout the project. Iterative implies that in the process of gathering information, testing and analyzing it the researcher comes closer to a model of the phenomenon studied. A general plan in form of an interwiew guideline with open ended questions is frequently used to support the interviewer (Rubin and Rubin, 1995, 43 - 47). The role of the interviewer is the one of an interested listener who wants to obtain as much information as possible. Generally, no closed questions should be used but for avoiding misunderstandings closed questions can be included to probe.

The interviewer has to be a good listener and subtly direct the flow of the conversation. The advantage of using a guide line is that the answers can be easily compared when analyzed; the open questions secure that the interviewee talks most and shares his knowledge and that the relevant topics are covered throughout the interview.

One of the main problems in qualitative interviewing is the influence of the interviewer on the interviewee. Depending on interests, personality, experience, and knowledge of the interviewers they will interpret the answers of the respondent. Statements that are wrong, statements in accordance with the interviewers' views or completely the opposite will stick more in the interviewers' memory (Tema-Lyn, 1999). The interviewers can solve this problem by making themselves aware of his attitudes and interpretation patterns (Maxwell, 1996, 90). A certain degree of sympathy should exist between interviewer and interviewee but should not result in not detecting or realizing wrong statements (Rubin and Rubin, 1995, 12).

The complete interviewing process includes seven stages. 1 *thematizing* (clarifying the purpose of the interviews), 2 *designing* the process through which the purpose is accomplished, 3 *interviewing*, 4 *transcribing* (creating a written text of the interviews), 5 *analyzing* (determining the meaning of the gathered material), 6 *verifying* (checking the reliability and validity of the material) and finally 7 *reporting*.

The steps following the actual interview, being transcribing, analyzing, verifying and reporting will be explained in the following sections.

2.1.4 Qualitative Data Processing and Analyzing

It is vital to make full and accurate notes of the interview and what went on. Even tape recorders and cameras cannot capture all that happened. If possible, one should take notes during the interview which is not always easy. When this is not possible the interviewer should write down the notes as soon as possible. The notes should include the information and an interpretation. A tape recorder can be used to support the transcription of the interview but the interviewer should always ask if the interviewee agrees with taping the conversation. The interviewer should never trust his memory more than necessary (Babbie, 1998, 293). The tape can help to avoid misinterpretation and errors in the notes, wrong memories can be detected (Maxwell, 1996, 89).

Some of the information can be anticipated before the study. Thus, it is possible to support the note making with standardized forms (Babbie, 1998, 294). The notes should be transcribed and typed on the same day they were taken. By doing so as many details as possible can be included and ideally be double checked with the tape that was recorded earlier.

All the answers have to be interpreted regarding who said it and in what context it was mentioned (Kepper, 1996, 58). There are various approaches for the interpretation of qualitative data. Most of those are used in sociology and are not applicable for market research. In the following paragraphs only the ones relevant for market research are discussed (Kepper, 1996, 59).

When making a *summary* the text is repeated in a shorter form that still includes the important statements. Through paraphrasing the unnecessary words are deleted and the text is paraphrased. This enables the researcher to compare the various statements and delete the ones that are repeated. In the last step the statements are bundled to reduce the data and facilitate interpretation (Weinhold-Stünzi, 1994).

When coding, the same ideas and statements of the different interviewees are put into groups and categories. The target is to find a coding guideline that allows for systemizing and structuring the data (Kepper, 1996, 61). The first step is to find some categories in order to code the text according to those. A statement can end up in more than one category. Factors that are identified as most important throughout the interviews can serve as categories. Those deal with important issues of the survey which cannot be made up prior to the survey. Irrelevant text can be deleted (Rubin and Rubin, 1995, 239). In some cases it makes sense to reduce the data within one category by summarizing it.

A comparison of statements within the categories and between categories is a good start for the interpretation of the data. Inconsistencies do not need to be consolidated but explanations for the different answers should be found. To summarize above discussion one can state that making categories and coding the data enables the researcher to focus on the relevant information.

Evaluation of qualitative research using measures like reliability, validity and objectivity as in quantitative research is only applicable with some restrictions.

Objectivity involves the reduction of subjective influence on the data. Every researcher should come to the same findings doing the same research. In the case of qualitative interviewing it is impossible to conclude with the same findings as the interviewer influences the results due to his personality and interpretation. Additionally, the interviewee will behave differently facing different interviewers.

Reliability demands stable measurement and a certain pattern, using the same design. Details can be found in Ruyter and Scholl (1998). Validity provides that the findings are representing what really was measured. This can be reached by an open, flexible and communicative survey (Kepper, 1996, 214). Among other measures are transparency and consistency. *"Transparency means that a reader of a qualitative research report is able to see the basic processes of data collection."* (Rubin and Rubin, 1995, 85)

To stay within the scope of the survey the interested reader is referred to Rubin and Rubin (1995, 85;87) for more details on transparency and consistency (Rubin and Rubin, 1995, 85) and the next chapter discusses how the theoretical background on qualitative research was applied in this research project.

2.2 Carrying out the Qualitative Research

Potential experts were identified and contacted via e-mail in October 2003. Nearly all the interviews took place in the interviewees' premises in October and November 2003 and took between 30-45 minutes.

The interviewees were experts in the mobile industry, in research, and consulting. The aim was to receive information from all players and researchers along the m-commerce value chain. Thus, researchers were identified through their primary field of research and managers due to their positions in companies dealing with mobile services.

Employees from network operators, consulting companies, advertising agencies, mobile marketing companies, location based services companies, Universities, the International Telecommunication Union, etc. were among the respondents. Due to the fact that the interviewees came from various cultural backgrounds insights into the European, Asian and Northern American industry was gained.

The following table shows the companies and Universities involved in the interviews, the interviewees position, the geographical background, and the competencies covered on the m-commerce value chain (see Chapter 3. for details on the m-commerce value chain).

Table 2: Group of Interviewees and their Background

Company	m-commerce Value Chain Competencies	Position of Interviewee	HQ of Company
WigeoGis	CP, MS	CEO	Austria
A.D. Little	All	Manager	France
Waseda University Japan	All	Professor	Japan
University of California Berkeley	All	Professor	USA
International Telecommunication Union	All	Policy Analyst	Switzerland
MobileMemoir LLC	CP, MS, MM	President	USA
12Snap	All	Senior Consultant	Germany
Metronet	MT, MS, MM, MI	CEO	Austria
t-systems	MS, MI	Member of Executive Board	Germany
Mobilkom	All	m-commerce Head of Dep.	Austria
IT Verlag	All	Editor	Austria
t-mobile	MM, MT, MS, MI	Executive Vice President	Germany
Saachi Saachi	CC, CP, CM	Chief Executive Officer	Each Country with rep.
One Connect	CP, MM, MT, MS, MI	Head of e-Business Development	Austria
Octane	MS, MI	Marketing Director	USA
Access	All	Research & Development	Japan
Webraska.com	All	Senior Manager	USA
NTT DoCoMo	All	CEO Research Laboratories	Japan
Booz Allen Hamilton	All	Principal	France
University of Western Australia	All	Professor	Australia

HQ=Head Quarter, CC=Content Creation, CP=Content Packaging, MM=Market Making, MT=Mobile Transport, MS=Mobile Services and Delivery Support, MI=Mobile Interface & Applications

After 20 interviews no more new findings were generated, a stage of 'theoretical saturation' was reached. Thus, no more interviews were conducted. In a qualitative survey not the number of equal answers but the richness of data generated is of relevance, therefore, the termination of the interview process was feasible.

A flexible guideline was used for the interview. The questions included in the guideline covered the areas that needed exploration according to the research problem and the deducted research questions. The guideline allowed the interviewer to go into more detail where necessary and receive high quality information from the interviewees. The guideline was adapted during the course of the qualitative research phase and included the following questions:

- Which factors lead to end-user adoption of mobile services?
- Which are the drivers of mobile service quality?
- What do you consider to be the most successful mobile services? Why?
- Is it possible to create value for the customer through mobile services?
- What could be determinants for perceived value in this context?
- What makes a loyal mobile service customer?
- What is the profile of a typical mobile service user?
- What trends can you identify in Europe, Asia and the USA?

Nearly all the interviews were conducted in a face to face interview; some, though, were completed through sending the guideline via e-mail and discussing the questions on the phone. The guideline should provide that all the relevant topics were covered in the interviews.

The interview was held by one interviewer, taped with the interviewees' permission and transcribed on the same day. The interviewer encouraged the interviewee to respond freely in form of an open conversation. Thus, yes/no answers could be avoided and open answers provide that the interviewee talks a lot. Generally there was a willingness to give away a lot of information. Some of the experts offered assistance in case questions occurred at a later time. The interviewees recommended some literature and Web sites containing additional information for the survey.

The following chapters present the findings of the explorative research starting with an introduction to m-commerce, the m-commerce value chain, m-commerce in different regions and a taxonomy of applications. The arguments are based on both literature review and the expert interviews.

3 M-COMMERCE VALUE CHAIN

The following sub-chapters present the findings of the explorative research and the literature review.

Recently mobile (m-) commerce has emerged fuelled by the immense success of voice communication via mobile phones and high penetration rates around the globe. This high penetration rate with mobile technology, such as mobile phones and personal digital assistants (PDAs), changes the way business is condcted.

Reviewing previous research Barnes (2002a) condenses that no single company in the m-commerce industry has what it takes to serve customers online demand. Hence, diverse inputs must be combined to create and deliver value. The inputs of diverse industries, only peripherally related in the past are needed (Schleuter and Shaw, 1997; Tapscott, 1995). Barnes (2002a) further stresses that various players have to cooperate. As a result, companies in telecommunications, computer hardware and software, entertainment, creative content, news distribution, and financial services have increased their opportunities by aligning competencies and assets via mergers and acquisitions, which leads to a consolidation of information-based industries (Symonds, 1999).

The following chapter provides an understanding of the new and unexplored field of m-commerce following Barnes' (2002a) value chain approach to cover all important aspects. The following chapter explores the nature and potential of mobile commerce in business-to-consumer markets.

3.1 The Basic Model

M-commerce involves key players in a chain of value-adding activities like any product or service. The chain terminates with the customer, which should increasingly be integrated into the value chain. Traditional value chain analysis by Porter could be used to unravel complexity (Porter and Millar, 1985). In the project at hand the value chain suggested by Barnes (2002) which was adopted from the European Commission (1996) and was also employed in other areas is used (Loebbecke, 2001; Schleuter and Shaw, 1997). Within this framework the the players, technologies and activities involved in m-commerce are analyzed (Barnes, 2002a).

The basic model consists of six main parts within two areas: (a) content, and (b) infrastructure and services.

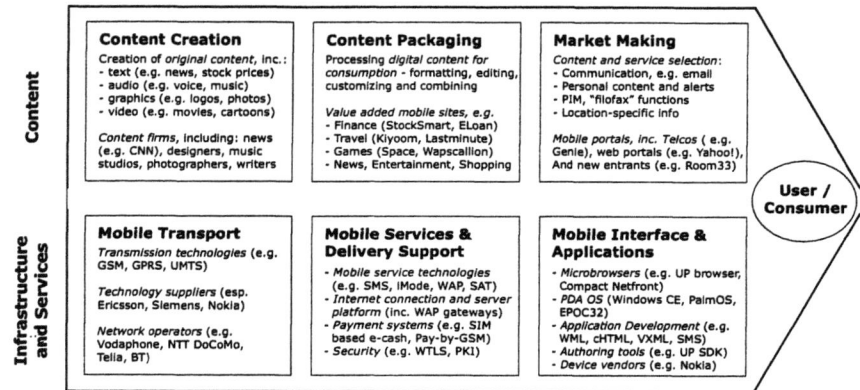

Figure 2: The M-Commerce Value Chain (Barnes, 2002a)

The six components of the m-commerce value chain have been illustrated above and will now be explored in more detail in the following chapters.

3.2 Infrastructure and Services

3.2.1 Mobile Transport

Mobile transport consists of the basic network involved in communications, including transmission, transportation and switching for voice and data. This includes major telecommunication players such as AT&T, NTT DoCoMo, Sonera, Vodaphone, Telia and BT. These enable, due to their infrastructure advantages in transport, movement along the value chain towards mobile services, delivery support, and market making (Müller-Veerse, 1999). The mobile network operators are responsible for billing and subscriber identification module (SIM) cards on the phone. In line with Barnes (2002a) the qualitative survey shows that they are in an ideal position to become portals and mobile service poviders and further expand into the field of mobile Internet.

An overview of some present and future transport technologies can be found in Table 3. It also includes recently introduced transmission technologies such as the Universal Mobile Telecommunications System (UMTS). The results of the study show that UMTS primarily brings bandwidth but what companies have to focus on is the improvement of the services offered to the clients instead of promoting technologies.

Equipment vendors and network operators are the main players involved in adding value in the transport element of the value chain.

Table 3: Key Mobile Network Technologies (Barnes, 2002a)

Standard	Description	Speed
GSM (Global System for Mobile Communication)	The prevailing mobile standard in Europe and most of the Asia-Pacific region – around half of the world's mobile phone users.	14.4 kbit/s
PCS (Personal Communications Services)	A standard based on Time Divisions Multiple Access (TDMA), which divides a frequency into time slots and gives users access to a time slot at regular intervals. TDMA is used in the US, central/south America and many other countries.	14.4 kbit/s
PDC (Personal Digital Cellular)	A standard used in Japan. Uses packet-data overlay on second-generation networks to achieve 'always on' data communication and a higher speed.	28.8 kbit/s
HSCSD (High Speed Circuit Switched Data)	A circuit switched protocol based on GSM. It is able to transmit data at around four times the speed of GSM by using four radio channels simultaneously. Some services were launched in late 1999 and early 2000.	57.6 kbit/s
GPRS (General Packet Radio Service)	A packet switched wireless protocol as defined in the GSM standard offering instant, 'always on' access to data networks. The speed will initially be less than the maximum burst: at first 43.2 kbit/s upstream and 14.4 kbit/s downstream rising to 56 kbit/s shortly afterwards.	115 kbit/s (burst)
EDGE (Enhanced Data rates for Global Evolution)	This is a higher bandwidth version of GPRS and an evolution of GSM. The high speeds will enable bandwidth-hungry multimedia applications. EDGE conveniently provides a migration path to UMTS by implementing necessary modulation changes. Planned service availability is for 2002.	384 kbit/s
IMT2000 (International Mobile Telecommunications)	This is a third generation (3G) standard. Three rival protocols have been developed: Universal Mobile Telephone System (UMTS) in Europe, Code Division Multiple Access (CDMA) 2000 in the US, and Wideband-CDMA in Japan. The development of the standard requires significant investment in infrastructure. Commercial availability of most services is predicted for 2002-3.	384 kbit/s – 2 Mbit/s

The experts pointed out that technology around the globe differs to a great extent. Europe chose the way of standardization with GSM and now recently implemented UMTS networks. In the US there is a trend towards GSM but they started off using CDMA 2000 and Japan using Wideband CDMA. Other Asian countries primarily use GSM protocols.

3.2.2 Mobile Interface and Applications

The nature of communication via the mobile device is very different compared to standard PC use, hence, the development and integration of an application interface for the user is essential. The key focus upon development has to be the users' needs not technological constraints (Barnes, 2002a). The results emphasize the companies' reluctance to involve customers' needs and wishes in the device development process. Chapter 3.2.2.1 deals with the mobile interface in more detail.

In the PDA and phone market the brand and model are the most important factors in purchase decision. Even the service and network providers are less important than the brand of the mobile (Hart, 2000). The qualitative survey revealed that sometimes customers can not even differentiate between mobile network provider and mobile phone manufacturer. Therefore, the mobile phone producers have a lot of power. They for instance decide which technologies will be incorporated in new handsets. The experts stressed personalization of the mobile phone to be one of the most important issues and motivators to buy a specific phone. According to them this trend is most evident among Asian and young users in general.

The main focus of the following chapter is on mobile phones as these are the most frequently used mobile devices and are highly relevant for this book.

3.2.2.1 Mobile Phones

The results show that the mobile phones are not only a telephone but an overall communication and entertainment tool (Haghirian, Dickinger et al., 2004). This trend from voice only to data transfer via the mobile device has been observed recently (Dickinger, Murphy et al., 2003). Mobile phones are more than a medium for exchanging voice/sound, they are about to establish themselves as an information terminal to exchange text messages, send and receive e-mail via the Internet and browse the web (Hashimoto, Komatsu et al., 2001; Mikami, Nakamura et al., 2001). Mobile phones are convenient communication tools and people can connect with anybody at anytime (Mikami, Nakamura et al., 2001).

Funk (2003) argues that the *"initial success of entertainment contents in 1999 caused manufacturers to introduce phones with color displays, polyphonic tones, cameras and Java programs"*. These functions are supported by other technological improvements like faster microprocessors, larger memory, and faster network speeds. According to Funk (2002), technologies are making the phone a portable entertainment player, a new marketing tool for retailers and manufacturers, a multi-channel shopping device, a navigation tool, a new type of ticket and money, and a new mobile Internet device. The following six new roles, functions or usages of mobile phones were mentioned by the interviewees, the arguments are supported by literature.

3.2.2.1.1 Phones as Portable Entertainment Players

The expert interviews revealed that young people primarily use the mobile phone as an entertainment device. Games, ring tones, screen savers, and other entertainment content is making the mobile phone a portable entertainment player. Faster network speeds, increased processing power, Java, and 3D rendering techniques reinforce this trend (Funk, 2003).

3.2.2.1.2 Phones as a New Marketing Tool for Retailers and Manufacturers

Mobile Marketing messages are used for branding, image campaigns, m-CRM, information services, mobile coupons, entertainment, location based services, product launches, push and pull based advertising etc. (Dickinger, Haghirian et al., 2003). Phones are a new marketing tool for retailers and manufacturers due to the lower cost and faster response time of the mobile Internet. At the same time, the experts stressed that marketers have to carefully use this tool for marketing as it is a private and intimate device and customers will not tolerate spam or other disturbances on their phone.

More than 100 retailers and manufacturers are using the mobile Internet to send discount coupons, conduct surveys, offer free samples, and improve their brand image with young people. Tsutaya Online, Japan's leading video retailer sent more than 100,000 coupons. Funk (2003) contends that a next step is to use the phone for loyalty programs and, thus, replace magnetic or paper cards.

3.2.2.1.3 Phones as a Multi-Channel Shopping Device

Funk (2003) mentions several technologies which are reinforcing an existing trend towards making the mobile phone a multi-channel shopping device. The small screens and keyboards make it difficult to search for products as convenient as with the fixed-line Internet. This is why most of the products purchased with a mobile phone are selected from personalized mail services that provide information on recent releases for a specific artist, genre, or author. The fastest growing segment of mobile shopping is buying products that are advertised in magazines together with a product code or URL, which can be dialed directly into the phone (Funk, 2003).

3.2.2.1.4 Phones as Portable Navigation Guides

The navigation market is one of the largest potential markets for mobile Internet services (Scharl, Dickinger et al., 2005). Improvements in the GPS function, larger displays, and 3D rendering techniques will enable mobile phones to become an important portable navigation guide for business and consumer applications (Funk, 2003). Current prototype development in the automotive industry focuses on traffic alerts in case of accidents, and on early-warning mechanisms to avoid potential dangers (Xu, 1999).

3.2.2.1.5 Phones as Tickets and Money

Some authors found an affinity of the financial services and insurance sector towards mobile commerce, due to developing electronic payment systems (Dickinger, Murphy et al., 2003; Kannan, Chang et al., 2001). Customers pay using SMS messages and are billed by their regular mobile phone provider. Since mobile devices are personal devices, the phone number identifies the individual payment via their cell phone. Kannan et al. (2001) also discuss the supplement to cash and credit cards in wireless commerce. Not only Japan but also European countries host an increasing number of pilot projects. Scandinavian consumers for instance purchase Coca Cola through vending machines featuring *"Dial-a-Coke"* (Johnston, 2000).

It is possible that phones used as tickets and money in the next few years will supplement cash continuing the move from physical to electronic money, starting with credit cards 50 years ago. The two methods discussed in Japan are infrared techniques and smart cards (Funk, 2003).

3.2.2.1.6 Phones as a Mobile Internet Device

The results show that Internet access and e-mail are among the most popular services. Mobile mail is widely used by business people and access to PC mail on the phone is one of the most popular services (Funk, 2003).

3.2.3 Mobile Service Technologies and Delivery Support

The development of platforms for mobile service delivery goes hand-in-hand with network standards. Interest groups ensure the establishment of platforms, among these interest groups is the UMTS Forum, the European Telecommunications Standards Institute and others.

Key mobile service technologies are given in the table below (Barnes, 2002a).

Table 4: Key Mobile Service Technologies (Barnes, 2002a)

Service	Description
SMS (Short Message Service)	Allows test messages of up to 160 characters to be sent to and from mobile handsets via a store-and-forward system. Although a large proportion of this is based on person-to-person communication and voicemail, other services such as news, stock prices and SMS chat are growing in popularity. Around 500 billion messages were sent in 2001.
MMS (Multimedia Message Service)	This is a new messaging service supporting graphics and audio currently on trial in Europe. It plans to build on the success of SMS.
CB (Cell Broadcast)	Not to be confused with citizen's band (CB) radio; this is another text messaging service. However, unlike SMS, CB provides a one-to-many broadcast facility that is ideal for push-based information services such as news feeds.
SAT (SIM Application Toolkit)	This allows applications to be sent via CB or SMS in order to update SIM cards, e.g. for downloading ring tones. Data security and integrity are standard features making it a popular choice for mobile banking. The WAP 2.0 standard will be compatible with SAT.
WAP (Wireless Application Protocol)	WAP is a universal standard for bringing Internet-based content and advanced value-added services to wireless devices such as phones and PDAs. In order to integrate as seamlessly as possible with the Web, WAP sites are hosted on Web servers and use the same transmission protocol as Web sites, which is hypertext transfer protocol (HTTP). The most important difference between Web and WAP sites is the application environment. Whereas a Web site is coded mainly using hypertext markup language (HTML), WAP sites use Wireless Markup Language (WML), based on eXtensible Markup Language (XML).
MExE (Mobile Station Application Execution Environment)	This standard is aimed at incorporation Java into the mobile phone providing full application programming. MExE is compatible with WAP but incorporates many other sophisticated services including voice recognition and positioning technology.
J2ME (Java 2 Micro Edition)	A version of the Java language designed for small devices. This is somewhat similar to MExE.
iMode (information mode)	iMode uses a variant of HTML for the provision of Web pages. iMode enabled Web sites utilize pages that are written in compact HTML (cHTML) – a subset of HTML 4.0 designed with regard to the restrictions of the wireless infrastructure.
iAppli (information application)	From January 2001, an upgraded version of iMode was provided in Japan to premium customers. The new service, iApply, is based on Java. Applications can be downloaded and stored, thereby eliminating the need to continually connect to a Web site. Further, constantly changing information is automatically updated at set times, e.g. stock prices or weather forecasts.
PDA Web clipping	This technology allows popular PDA devices, such as PALM and Handspring, to access dynamic and updated HTML content via a modem. Web clipping is used in combination with applications stored on the device.
PDA (Syncing)	This allows PDAs to store or cache content without the use of a wireless modem. Content is updated when the user synchronizes ('syncs') or connects their PDA to the Internet via computer connection.

SMS, the most frequently used and widespread messaging service technology, is explained in more detail in the following paragraphs.

3.2.3.1 Messaging Services

The interviewees emphasized that SMS is the most successful mobile service technology. The main reasons are that it is easy to use, the user knows the cost of each message and trusts the service.

Text Messaging, as it is called in the UK or Short Message Service (SMS) in other European countries, the US, and Australia, lets users send and receive text messages via their cell phones. According to the GSM Association, users send more than 10 billion SMS messages worldwide each month (Cohn, 2001). Normally messages arrive within minutes, although there is no guaranteed delivery time and longer delays may occur.

With only 160 characters available, a major problem facing mobile marketers is designing attractive text messages. Emerging technologies such as the Multimedia Messaging Service (MMS), which enhances messages by incorporating sound, images, and other rich content, should overcome this limitation (OMA, 2002).

A usage analysis of SMS showed that the age groups from 15 to 24 are heavy SMS users. They can be reached easily, when SMS is used for marketing services which is often difficult via other media (Puca, 2001).

Nowadays, the majority of the mobile consumers also subscribe to the mobile Internet and e-mail service. Hashimoto (2001) claims that messaging services are the most frequently used function of cellular phones. For almost all of the interviewees, messaging communication – i.e. communication via SMS or mobile e-mail – prevailed over voice communication – i.e. communication by making phone calls.

Zobel (2001) mentions a survey which revealed that mobile phone consumers' usage of the phone for making calls makes up only 40%, while mobile Internet and mail account for the major part (Zobel, 2001). The rapid expansion of cell phone usage has changed the way people communicate (NTT DoCoMo, 2002). Nomura (2003) speaks of a demand shift from telephone calls to mails. As mentioned above, nowadays, the majority of mobile consumers also subscribe to the mobile Internet and e-mail services, which is particularly true for the Asian market. More than 38 million out of NTT DoCoMo's 46 million mobile phone customers have subscribed to i-mode. Messaging services are the most frequently used function of cellular phones (Hashimoto, Komatsu et al., 2001). The key factors and motivations supporting a shift from voice to data transfer via the mobile device are:

- Convenient Usage

Convenience and cost performance were mentioned by all interviewees as drivers for messaging communication. Another reason is that messaging communication can be used all the time; even when the other person is busy at work. Sending an e-mail is a more discreet way of contacting people and gives the receiver of the message more freedom to think and respond to it.

- Cost Performance

Mikami et al. (2001) found that when asking for the motive to start using mobile mail: 52.7% of the respondents said that it was cheap/economic and for 50.7% the motive was that mobile mail can be used regardless of time and place. However, it should also be taken into consideration that there are completely different usage motives too. 50% said that they started using mobile mail because their friends or boyfriend/girlfriend are using it and 33.3% argued that it seemed to be interesting to try (Mikami, Nakamura et al., 2001). Similar results were also found by Hashimoto et al. (2001).

- Personal Contact Possibilities

Peer relationships are a main concern of adolescents. Etiquette in Japan is different to other countries, voice communication for example is not used while riding a train (Mikami, Nakamura et al., 2001). Especially those who have to commute to the office or to school usually spend daily many hours on a train. A survey of elementary and junior high school students conducted by NTT DoCoMo in Japan found a strong connection between mobile phone ownership and a student's way to school. One of the main arguments is that the longer the way to school the more important the phone for the young person (NTT DoCoMo, 2001a).

Occasionally messages are sent without any particular reason by young people. For instance, friends might ask each other about their whereabouts or simply want to say *"hello"* or *"how are you?"*. In this case, people simply do not consider the content of what they want to say important or urgent enough to make a phone call. An interesting phenomenon in this context is the so-called consummatory usage of mobile phones or consummatory communication (Mikami, Nakamura et al., 2001). The only purpose of the communication in this case is the usage for its own sake and becomes the end in itself. This means that for private users, the mobile phone is not a tool or instrument to achieve a goal/purpose any more but using it simply becomes an end in itself. Of course, this is true for phone calls as well as mobile mails.

3.2.3.2 Security and Privacy

One major concern in delivery support is security as the development of common security standards is unsolved (Manchester, 2000). However, WAP 1.3 introduces standards that use digital certificates, certificate authorities, strong asymmetric encryption and digital signatures to ensure integrity, privacy, authenticity, and non-repudiation (Barnes, 2002a; Stein, 1998).

Mobile spam is a potential danger companies have to be aware of. Corporate policies must consider legalities such as electronic signatures, electronic contracts, and conditions for sending SMS messages. Seven experts who had used SMS campaigns welcomed European government and industry initiatives to restrict unsolicited SMS. They argued that sending unsolicited messages hurts the mobile commerce industry.

According to the experts, advertisers should have permission and convince consumers to *"opt-in"* before sending advertisements. A simple registration ensures sending relevant messages to an interested audience (Petty, 2000). Unsolicited messages, commonly known as spam (Hinde, 2003; Wales, 2003), stifle user acceptance (Golem.de, 2002) – particularly as mobile phones cannot distinguish between spam and genuine communication automatically. Unwanted messages are illegal in some countries (Stratil and Weissenburger, 2000) and annoy consumers regardless of the medium (e.g., fax, telephone, electronic mail, or mobile devices). All experts cited fear of spam as the strongest negative influence on consumer attitudes towards mobile services.

Changing one's mobile phone number is more difficult than changing e-mail addresses provided by free services such as *Yahoo!* or *Hotmail*. New regulations allowing people to keep their phone numbers when switching cellular carriers (CNN, 2003) may reinforce fears of unwanted messages and misuse of personal data, thereby keeping consumers from registering for SMS-based information services.

3.3 Content

3.3.1 Content Creation

Generally content creation and delivery is the same in m-commerce and e-commerce (Choi, Stahl et al., 1997) but because of the size of the devices the specific format is different. Typically such mobile content can consist of (Barnes, 2002a):

- Text
- Audio
- Graphics
- Video

This electronic content can be consumed repetitively by the same or by different consumers. In addition it can easily be modified and fast and cheaply reproduced. The attributes of online delivered content (ODC) are transmutability, indestructibility/non-subtractivity and reproducibility (Loebbecke, 2001).

Barnes (2002a) summarizes some important issues that have to be taken care of when digital content for the mobile device is created (Choi, Stahl et al., 1997; Loebbecke, 2001). Among others these include interactivity and customization, time-dependence, use frequency and the operational format (executable vs. fixed document).

3.3.2 Content Packaging

To make mobile content consumable it must be edited, customized or combined, before sending it to the customer. At this stage of the value chain, value is added by transforming the data into the most convenient form for consumption (Barnes, 2002a). Some critic remarks of the interviewees emphasized the importance of transforming web content into a format that is appealing on the small sized display of a mobile device. There is a huge variety of mobile content which includes (Barnes, 2002a):

- *Sports*
- *Online Games*
- *Finance*
- *Entertainment*
- *News*
- *Shopping*
- *Travel*

At the moment an integration of different media makes the customer aware of mobile services. Web pages often serve as a platform to register for mobile services.

Sending data via text messaging is time-consuming. Web-based information systems, by contrast, offer easier registration and more accessible interfaces due to a computer's larger keyboard and higher screen resolution. Thus, consumers may prefer to share their interests, desired content, and number and timing of messages via their home or office PC. A good example of gathering user preferences, interests, and permissions via the Web and the popularity of multi-channel strategies is MindMatics' RedAlertz (http://www.redalertz.co.uk/).

Consumer trends, especially increased Internet usage, influence the evolution of mobile marketing. Integration with Web-based information systems is crucial since many customers subscribe (opt-in) to SMS campaigns via corporate Web sites, which are often the first and primary point of contact. In addition to that, the acceptance of mobile services is probably higher by Web users than by the overall population.

Several companies – e.g., Flytxt's FXTrinity, Mindmatic's Wireless Interactive Box, UCP's Mobile Media Platform, and the Chordiant Marketing Director Suite (Chordiant, 2003) illustrated in below figure – already integrate mobile and Web-based channels.

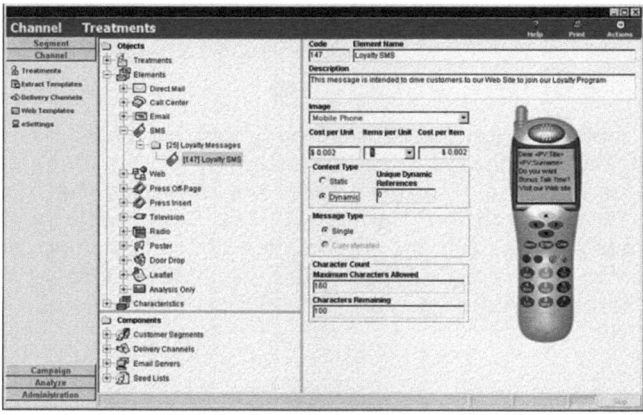

Figure 3: User Interface of the Chordiant 5 Mobile Marketing Director

3.3.3 Market Making

Mobile portals (or m-portals) are the most important business-to-consumer market makers on the mobile Internet. Mobile portals cumulate services and information (Ticoll, Lowy et al., 1998) or serve as intermediaries (Chircu and Kauffman, 2001). Hence, they have a powerful role in the access of mobile information (Barnes, 2002a). The aim of portals is the provision of content fulfilling customers' needs. These needs can include (Barnes, 2002a):

- *Communication*
- *Personalized content and alerts*
- *Personal information management*
- *Location-specific information*

The probably best example for mobile portals is the Japanese network operator NTT DoCoMo. The provision of content for iMode is controlled through their portal page (Nakada, 2001).

Now that the key elements of the m-commerce value chain have been discussed and analyzed, the next and logical question is how the value chain can be further leveraged to create a profitable partnership with customers.

3.4 The Mobile User

In the last years an amazing number of new functions has been developed and added to cellular phones. Besides, with the technology becoming more advanced and sophisticated, screens became larger – with more colors and better resolutions – transmission rates higher and the sound quality much better. Obviously, these changes and developments in communication have influenceg people's lifestyles and habits. This reciprocal process of mutual influence, change, and development might also lead to a transformation of attitudes, understanding, and perceptions of consumers. Geser (2002) states, that the mobile phone has an influence on social behavior. It influences the individual, the way interpersonal interaction is conducted and has implications on face-to-face gatherings. The mobile device has changed the way people organize their social relationships both at work and at home (Haddon, 1997).

Experts in the industry contend that *"in terms of 3G implementation, Japan is two or three years ahead of Europe"* and that European companies should *"take the opportunity to learn from the Japanese experience"* (DeZoysa, 2002). Thus, Japanese consumer surveys can give an overview on usage behavior and insights into the social phenomenon of the mobile device.

An ever-rising importance of mobile phones in every day life can be observed. In the NTT DoCoMo Report on current trends in mobile phone usage among adolescents, 45% said that a mobile phone is *"essential in my life"*, in terms of overall impression and 46% said that *"it is not essential in my life, but I want one"* (NTT DoCoMo, 2001a). Another NTT DoCoMo (2001b) report found that *"mobile phones have become an essential part of people's daily lives"*, with nearly 70% of the respondents considering mobile phones *"essential for enjoying their lives"*. Mikami et al. (2001) of the Institute of Socio-Information and Communication Studies at the Tokyo University concluded that the cell phone spread as an everyday medium. Regardless of one's own view of media it became a media indispensable to use for communication at work as well as for communication between the young generation (Mikami, Nakamura et al., 2001).

On the one hand, mobile phone users stress the convenience of the cell phone – e.g. that they are *"not tied down by time or location, can act freely"* – and that it gives them a feeling of reassurance (Mikami, Nakamura et al., 2001; NTT DoCoMo, 2000; NTT DoCoMo, 2001a). On the other hand, there is a growing number of users who complain about a certain over-dependence on mobile phones. In fact, 52% of those surveyed by NTT DoCoMo in 2000 answered that they *"feel lost without it"* (NTT DoCoMo, 2000). 32% of the respondents to the survey conducted by Mikami et al. (2001), stated that they feel uneasy if they do not carry the cell phone with them all the time (NTT DoCoMo, 2001a) and that especially people in their 20s become more and more nervous if they cannot reach somebody or cannot be reached themselves (Mikami, Nakamura et al., 2001). Hashimoto (2002) identified very similar trends and already speaks of a *"cellular-phone dependency"*. Many of the interviewees also complained about a certain loss of privacy and the fact that one is constantly within call when carrying a mobile phone. One interviewee even stated that she has the feeling that the *"phone is part of the body already"*.

The explorative survey showed that mobile applications are valuable and useful for certain user segments, mainly people between 16 and 35. Particularly students and frequent travelers can profit from mobile services. According to experts the adoption curve is promising in those segments. For other user segments, it is still not clear what the adoption curve will look like.

According to the experts the perception of usefulness and value differs between segments. Younger users have a different set of priorities as opposed to traveling or business users. Furthermore, one key factor will be the degree of personalization of the messages. Overall, before developing a service for a particular segment, businesses have to look at the factors motivating people to use a service, at the value propositions, pricing models and the marketing strategy. The more focused a service is the more successful it will be within its target group; the downside is that the company is inherently limiting itself to a small group.

Pupils and students have unique requirements. For one, regulations surround collecting profile information about kids, so customized information services may not do well with youth. However, youth generally have a lot of time at their hands which they need to fill with diversionary entertainment. They ride buses to school, they never drive, they have boring classes at school, etc. Thus, community and gaming applications provide value as a way to pass the time. Young people even express themselves through the mobile phone.

The other major user group is business people and travelers who want to save time and, thus, use a mobile device. The device supports them in every day life, transmitted content is relevant and useful. Services like the mobile Internet are particularly important for this user segment. The results of the explorative research show that users are spoiled by the Internet. They are used to receive free content but via the mobile device they must pay for services they are used to get for free via the Web.

According to the experts the main drivers for mobile service adoption are clear value propositions and high quality of the service. The main influencing factors for quality are efficiency enhanced by the service, it has to be easy to use, the usage should be fun, and trust into the technology and reliability of the service provider are essential.

The following chapter gives some more detailed insights into mobile services. Then the theoretic body of service quality and diffusion of innovations is examined since the main influencing factors of mobile service use have their theoretic roots in those domains.

4 Mobile Services

The literature review and survey found six characteristics to describe successful mobile services. First the characteristics are presented, then, some groups of mobile services are described. The number of services explained is not exhaustive, the ones considered most important by experts and that raised the attention of scholars leading to literature on the services are discussed below.

4.1 Characteristics of Mobile Services

Services in the mobile commerce area have some advantages over conventional e-commerce applications. This section first introduces the most obvious characteristics.

Ubiquity is the most obvious advantage of the mobile device (Balasubramanian, Peterson et al., 2002). Mobile devices fulfill the need for real time information and communication at a level the desktop PC is unable to provide. The factor of ubiquity should be realized in a multi network environment enabling seamless access to services regardless of the network provider.

Location based services may be more common with the next generation of mobile phones carrying a GPS chip allowing for precise locations. It is even emaginable that when clients pass a specific shop they could receive an alert with special offers. Also the time sensitiveness of this approach is critical since they should receive the message when in front of the shop and not half an hour later. The experts question if the necessity of location based services is over estimated. People are normally in locations they know. If they are abroad concerns with regard to roaming come up. Furthermore, some degree of usage is necessary to make services profitable for companies.

Push is the predominant wireless delivery method. It saves time and money compared to surfing the Internet via WAP, but the information should be relevant to the receiver (Quah and Lim, 2002). Yunos et al. (2003) state that most users view wireless services pushed to their mobile device as intrusive and unwelcome.

The proposed taxonomy below distinguishes m-services along three dimensions, based upon the extent an application is: i) location-sensitive, ii) time-critical and iii) user- or provider-initiated (Balasubramanian, Peterson et al., 2002).

Table 5: Taxonomy of Mobile Services Adapted from (Balasubramanian, Peterson et al., 2002; Dickinger, Heinzmann et al., 2004)

Dimension I	Dimension II	Dimension III	Examples
Location Sensitive	Time critical	User initiated	Up-to date information, mobile market research.
		Provider initiated	GPS Systems in sports, route planner.
	Time uncritical	User initiated	Information services, booking of hotels, restaurants and concert tickets.
		Provider initiated	Advertisements of restaurants or shops one passes.
Location In-Sensitive	Time critical	User initiated	Weather forecast on demand, virtual discussion forums.
		Provider initiated	Warning systems, automatic updates of flight time change.
	Time uncritical	User initiated	Download of information and data, city maps, dictionaries etc.
		Provider initiated	Remote update of data, automatic downloads.

Group 1 applications provide timely information that relates to a receiver's surrounding (Balasubramanian, Peterson et al., 2002). These applications are particularly interesting for integration into users' every day tasks. From a managerial perspective, Group 1 applications can increase system efficiency and monitor the performance of individual mobile units (e.g. truck driver and engine performance parameters).

Group 2 applications differ from Group 1 applications as the product or service provider typically initiates and controls the application.

Group 3 applications relate to on-demand provision of information and account for the recipient's location. Examples with Global Positioning Systems (GPS) include locating one's position on a map when traveling, route directions, and information about the physical environment. Group 3 illustrations include pushing messages to mobile devices with information about smog, ozone ratings and pollution. Another example is directions to a restaurant, hotel, or point of interest.

Group 4 applications resemble Group 1 applications, except that they are less time-critical. A prominent example is John Deere's satellite-based yield mapping system, which lets farmers collect data about crops, field conditions and harvesting via a mobile device. Gaining specific information about different locations on the field, the farmer can treat each location with the appropriate fertilizers, pesticides and seeds (Deree&Company, 2004).

Group 5 and Group 6 are time-critical applications whereby the location of the user is less relevant. The user initiates Group 5 applications and pulls the information while the company or organizations push the information to the users with Group 6 applications.

Group 7 applications relate to receiver-initiated access to data via the mobile terminal. Users can access text at a satisfactory level but high data volumes limit receiving picture and sound. Generally, group 7 applications provide users with convenient access to information or facilitate other tasks.

Group 8 services relate to modifying, updating and configuring software at remote locations (Balasubramanian, Peterson et al., 2002). For example, earth-based signals moved a NASA telescope away from a meteor shower in 2001.

The proposed taxonomy describes a landscape for m- services, serves as a platform for further conceptualization of issues related to the mobile environment, and encourages practitioners to probe existing and potential services systematically.

The following section illustrates different types of services with regard to their content.

4.2 Types of Mobile Services

First information services are introduced, then transaction services, mobile entertainment followed by mobile communication and finally by mobile marketing services. The list of services is not exhaustive but rather intends to give an overview and examples on the mobile services landscape. In this research only end-user services (B-to-C or C-to-B) are of interest.

4.2.1 Information Services

Mobile data applications deliver content and enable transactions for users on the move, often drawing on the World Wide Web for content and the Internet as an enabling technology for Wireless Application Protocol (WAP) and i-mode (Gilbert and Kendall, 2003). WAP, mainly used in Europe and the Japanese i-mode are protocols that let users access information instantly via a mobile device. The majority of messages use the SMS format, which sends up to 160 text characters or little pictures to and from mobile devices. Commercial uses of SMS include weather reports, games, stock quotes, sports results and basic information (Kalluviayil, 2001; Massoud and Gupta, 2003).

Due to its popularity, SMS should be the primary mobile technology for information services such as sports, news, stock prices, weather, and e-mail notification until 2005 (Powell and Vu, 2002). Massoud and Gupta (2003) found that 30% of mobile users have experience with information services (finding weather, sports results) and 27% with location-based services (street navigation). The top two mobile services that interest consumers are entertainment at 88% and information at 66% (Massoud and Gupta, 2003).

Businesses deliver these services via push and pull models. Push messages are unsolicited, usually free informal SMS alerts from mobile network providers to subscribed users. Pull messages promote free or inexpensive information such as sport results, traffic reports or weather forecasts that the customer requested (Schreiber, 2000).

This category includes services that provide end users with news, updates and alerts. Essential for this type of service is that it will only poof successful if the information provided is highly personalized. The content has to be of high quality and tailored to the users' specific information requirements.

Information of the favorite football team or on the weather in a certain area could be of interest for instance. Popular services are weather forecasts, the horoscope, time tables of public transport, etc.

According to the experts information services are also provided via voice and could make people try out mobile services. Traffic alerts via SMS for instance work well (Dickinger, Heinzmann et al., 2004) and are a welcome alternative to the more expensive global positioning system.

4.2.2 Entertainment Services

A Siemens Survey and Gartner Research found that entertainment applications especially appeal to the "generation @" segment – young World Wide Web users between 12 and 16 years old (Hartmann and Büppelmann, 2001). Young people are heavy users of SMS services but often fail to realize how much text messaging and voicemail cost. The Communications Law Center in Australia found that a quarter of young people have difficulties paying their mobile phone bills and frequently end up in debt (Lowe, 2003).

Entertainment services can increase customer loyalty and add value for the customer. As most people have a natural playfulness, providing games and prizes via text messaging yields high participation, noted a mobile marketing expert. Sending games and prizes to the customer's cell phone is a fun way to attract and keep customers.

Warner Brothers Movie World in Germany, for example, invited customers to send a certain message to three friends as quickly as possible, asking them to forward the message to Warner Brothers. The first five teams to complete the cycle received free tickets to Warner Brothers' Movie World entertainment park (MindMatics, 2001).

Entertainment services have been a huge success in Japan with 68% of site access resulting from those (Devine and Holmqvist, 2001). The mobile phone became the ultimate gadget to kill time. For some people like long-hour commuting Japanese, it is an essential feature. Japanese mobile handsets evolve for this simple reason for years.

Mobile games have to grow increasing complex to satisfy the customers' demands. Interactive and location based games have been recently introduced and grow in popularity. Games are particularly important for adolescents. They have a relatively high disposable income and are willing to spend money on mobile games and are keen on personalizing their mobile device through ring tones, wall papers and individual covers.

One special field in the entertainment industry is music downloads. As opposed to the Internet the customers are used to pay for services when using the phone. Payment for music downloads will be increasingly tolerated via the mobile device.

Experts state that one-time downloads such as games and ring tones have healthy distribution channels and revenue models, and thus, enjoy the most success. Mobile users can easily purchase games and ring tones from their operators, from websites, and from mobile phone stores. Furthermore, users understand the value of a one-time purchase for a game or ring tone a lot better than they understand the value of a mobile information service.

In short, ring tones are so popular because of their simplicity, availability and ease of installation. Even more important, the ring tone enables the users to personalize their mobile device. Personalization of the device by choosing different ring tones, wallpapers and covers is one of the main issues when it comes to Asian users but is observed in Europe as well.

Music companies have the opportunity to sell ring tones of songs in the charts and earn money through that channel and extract revenue from that emerging market. According to experts music companies do not want to make the same mistake as with downloads of music from the Internet where they miss out on profit.

Gambling can be regarded as part of the entertainment services. It includes casino type games as well as betting on real-time sporting events. Thanks to the access via mobile phone the player remains anonymous. Mobile Lotto has been introduced in Austria as well and appears to be a successful service.

4.2.3 Transaction Services

The development of mobile payment systems helps explain (Kannan, Chang et al., 2001) the affinity of finance and insurance towards mobile commerce. Kannan et al. (2001) describe how mobile commerce supplements cash and credit cards – e.g., paying via SMS messages or using electronic wallets linked to pre- or postpaid services. Contactless smart cards facilitate transactions (Durix, 2003) by automatically connecting with public terminals while the customer is in range. Instead of SMS, proximity transactions typically rely on infrared signals, radio transmission or Bluetooth (Durix, 2003).

Pilot projects illustrate the feasibility of mobile payment. Austrians use PayBox to obtain parking permits or tickets for cultural events (Paybox.net, 2002), Scandinavians purchase Coca Cola through *"Dial-a-Coke"* vending machines (Johnston, 2000), and South Koreans substitute train tickets with their mobile phones when passing through electronic turnstiles (Durix, 2003). In Finland, Sonera's Pay-by-GSM enables the user to dial a number to receive a charge to a prepaid phone or for a deduction from a mobile account.

According to the experts mobile tickets are a cheap and convenient alternative to traditional ticketing. Cinema tickets and concert tickets can already be bought via the mobile phone. It is quite convenient not to queue to get ones tickets.

One further transaction service is mobile coupons. Companies can send coupons to cell phones via SMS. This mobile couponing offers at least three advantages: targeting based on customer cell phone numbers; time sensitivity, e.g. receiving a 20% discount on purchases immediately after entering a shop; and efficient handling by scanning the coupon's bar-code at the cash desk.

Raskino (2001) predicts that consumers will use mobile coupons 300 times more often than ordinary paper coupons. Customers keep their cell phone with them and consequently the coupon too.

M-Parking, one of the most successful mobile commerce service in Austria will be further investigated in the quantitative part of this book. M-parking was introduced by the biggest mobile network provider *Mobilkom* and *Siemens Business Services* in 2002. About 50,000 car drivers now pay their parking fee via their mobile phone in six Austrian cities.

On a web platform the drivers register their mobile phone number and car license plate. Credit for the parking fee is purchased either by credit card or a direct debit order. In order to pay the parking fee, the drivers send an SMS to a service number indicating how long they intend to park. The user receives a reminder SMS a couple of minutes before the parking period expires. They then decide whether to get back to the car or to extend the parking period by sending another text message. Similary to ticketing the user does not receive a physical good but a sort of good directly stored on the mobile phone which is the parking fee.

For all transaction services security and trust in the provider are the most important factors for end users to rely on a service.

4.2.4 Communication Services

Apart from voice traffic, SMS has so far been the most successful means of mobile data communication. As technology evolves new services will be surfacing. MMS for instance has already been introduced and mobile e-mail is gaining market share.

As mentioned earlier messaging services are the most popular communication services. One messaging service is Blackberry, first introduced in Austria by t-mobile. The mobile device supports one major function which is receiving of and responding to e-mails. The Blackberry, a very specialized device serves particularly this need.

Push to Talk (PTT), recently introduced in Austria, is a form of cellular communication that allows users to engage in immediate communication with one or more users. It is providing a *"walkie-talkie"*-type of communication. Rather than being a replacement of long, interactive communication, PTT is best suited for quick communication among end-users. PTT is provided in half-duplex mode e.g., transmission occurs in both directions, but not at the same time – each party must wait to speak. Due to the inability to interrupt it lends itself to quick exchanges of information.

4.2.5 Mobile Marketing Services

The following chapter introduces the new concept of mobile marketing. A definition of mobile marketing is followed by a presentation of mobile marketing opportunities.

The American Marketing Association's definition of marketing management provides a basis for defining mobile marketing:

*"Marketing management is the process of planning and executing the conception, pricing, promotion and distribution of goods, services, and ideas to create exchanges that satisfy individual and org*anizational goals" (AMA, 1985, 1).

This definition implies temporal and spatial separation of buyers and sellers as well as sequential marketing stages. These boundaries and distinctions, though, blur as mobile devices extend traditional marketing's time-space paradigm (Watson, Pitt et al., 2002).

Cell phones, free from traditional land-based Internet connections, amplify main arguments of e-commerce, location independence and ubiquity (Anckar and D'Incau, 2002; Balasubramanian, Peterson et al.,

2002; Watson, Pitt et al., 2002). At the same time, the importance of location, time and personalization in mobile marketing is reinforced by consumers who increasingly expect tailored and location based services (Watson, Berthon et al., 2000).

One-to-one Marketing, addressing customers individually, is well established in marketing and plays a central role in Customer Relationship Management (Kotler, Jain et al., 2002; Newell, 2000; Peppers, Rogers et al., 1999). As with other forms of digital marketing, mobile media incorporate interactivity and transcend traditional communication, allowing for one-to-one, many-to-many and mass communication models (Barwise and Strong, 2002; Hoffman and Novak, 1996; Jee and Lee, 2002). Although it can be expensive to learn individual customer interests, customized information treats each individual uniquely (Watson, Pitt et al., 2002).

The two-way information flow, sending and receiving messages, enables better service and feedback between companies and consumers (Jones, 2001). Text messages are also a popular instrument for interpersonal communication (Döring, 2002). Cell phones let users of all ages easily maintain business and social contacts, although research found that entry barriers to using SMS were high for teachers and other adult authority figures (Thiele and Liess, 2000).

Summarizing the discussion above, mobile marketing provides consumers with personalized information based on the time of day, their location, and their expressed preferences. Using the Global System for Mobile communication (GSM) technology and communication via Short Message Service (SMS) the customer can easily answer via SMS, interacting with the marketer.

4.2.5.1 Mobile Branding

Mobile marketing is a branding tool. For over a century, branding efforts have attempted to link images and emotions with a brand in order to gain a competitive edge beyond utilitarian differences (Alderson, 1965; Borden, 1942). Brands, usually a company's most stable asset (Clifton, 2002) and a fundamental tenet of business success, simplify consumers choice with a brand name that links closely to a product category (Rubinstein and Griffiths, 2001). For example, when most people think of fast food, McDonald's comes to mind. Barwise et al. (2002) posit that trusted brands are more important in the virtual world where they influence online purchases, generate customer loyalty (Clifton, 2002), and attract customers to their Web sites (Hanson, 2000). This 'virtual branding' effect may apply to SMS as well.

Wella, a leading seller of hair cosmetics and fragrances in over 150 countries, conducted a campaign that sent a message with a kiss to all their clients that gave permission to receive SMS messages from Wella.

The customers liked the Wella kiss so much that they forwarded it to their friends. This viral impact created a high effect for a low cost (Godin, 2001). Wella paid for text messages sent to the opt-in clients but paid nothing for the messages passed on to friends (12Snap, 2001).

4.2.5.2 Mobile CRM

Text messaging supports Customer Relationship activities such as receiving free newsletters, pictures, ring tones, bonus points and coupons after joining a customer program. Mobile network operators plan to use mobile marketing for customer relationships by sending their clients SMS information on where to get cheap pre-paid phone cards when their credits are running low.

One expert said that his cell phone company plans to use mobile marketing for customer relationship management by *"sending SMS based reminders if clients do not pay their bills on time. This kind of reminder is more effective and less expensive for us as operators. Of course the legal consequences need to be checked first."*

4.2.5.3 Mobile Advertising

The convergence of the Internet and wireless telephony serves as a new platform for advertising and is, potentially much stronger than the wired Internet. There are guidelines for successful communication but need to be adapted for the wireless medium (Barnes 2002).

The characteristics of the wireless devices (very personal, timely, location based) bear a high potential for advertising (Kannan, Chang et al., 2001).

Mobile advertising can be effected in two ways, in either a push or pull model. Push advertising sends messages, usually via an SMS alert, that the customer has not specifically requested. Pull advertising adds messages, usually promoting free information (e.g., traffic reports or weather forecasts), to browsed content or information the customer requested (Schreiber, 2000). These two main categories are illustrated in the figure below followed by an in depth discussion.

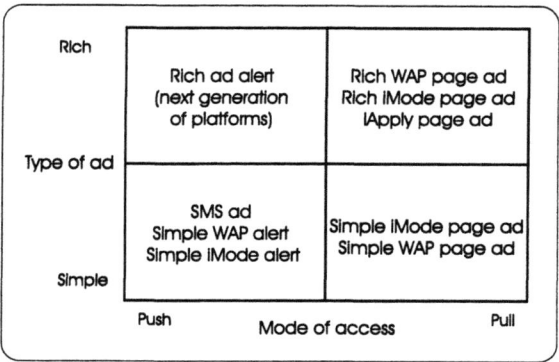

Figure 4: Categorization of Wireless Advertising with Examples (Barnes, 2002b)

Wireless Push advertising is currently the biggest form of wireless advertising due to the predominant usage of SMS, as mentioned earlier.

The idea of sending messages directly to an individual's private phone raises legislative concerns. It is obvious that push marketing should only be allowed for companies that have their customer's permission.

As a consequence 'opt-in' schemes become popular (Barnes 2002). The idea of opt-in schemes is that the user agrees to receive advertising before anything is sent and has the opportunity to change preferences or stop messages at any time. If an opt-in policy is followed some customers may even appreciate receiven advertising messages (Ericsson, 2000; Quios/Engage, 2000). Among the reasons for the positive attitudes to SMS advertising are (Barnes 2002):

- *Content value*. People were willing to receive advertising as long as their received some sort of compensation.
- *Immersive content*. SMS messaging has an interactive quality attracting young users.
- *Ad pertinence*. Now there is little junk messaging or spam. Advertising is not very common and thus, pertinent.
- *'Wow' factor*. Since there have not been too many SMS advertisements they maintain a degree of surprise.
- *Viral marketing*. Resulting from the high impact of advertisements and usage of mobiles for social contacts mobile advertising can have a viral impact.
- *Personal context*. The personalized services are particularly favored.

Further expansion of this type of advertising may change customers perceptions. A result could be that the user even starts to be annoyed (Hamblen, 2000).

At the moment push-based advertising is in the lower left quadrant of the above figure but this will change as new technologies emerge.

In *wireless pull advertising* any browsing enabled wireless platform can be used. The most important aspect of pull advertising is targeting to provide the user with relevant messages leading to positive response (Lot21, 2001). Considering the expensive connect time and small displays a positive user perception is imperative. One approach to overcome this problem is by positioning wireless advertising as additional content (Barnes 2002).

As opposed to the wired Internet, wireless advertising should reach the customer when they want to receive it, and thus, add value. Consequently, the line between advertising and service becomes blurred (Katz-Stone, 2001). Most WAP pull ads are simple in nature. They are on the bottom right quadrant of above figure. Types of ads in that context are as follows (Barnes 2002):

- Simple text ads
- Rich ads
- Interstitial ads

4.2.5.4 Mobile Market Research

Some companies carry out market research via the mobile device. They are working on building a panel of people willing to participate in "wireless market research". The initial difficulty is entering a URL on a browser since they have to be very short and should contain as few unusual characters as possible because people simply can not figure out how to enter them into their phone.

Feedback has been positive with panelists seeming to enjoy the experience and in some cases even asking for more or longer surveys. In a typical market research panel for surveys conducted over the web, a panelist will take 3-4 surveys and then they are tired of the experience. The main advantage is that customers can be reached during the consummation experience, e.g. on a holiday, and can give up to date insights into perceptions and experiences.

5 THEORETICAL AND METHODOLOGICAL FRAMEWORK

The following section suggests theoretical frameworks relevant for this survey. Among those are diffusion and adoption theories and service quality followed by the methodological foundations of the research project.

5.1 Relevance of the Theoretical Framework

The first two sections on adoption and service quality introduce a number of concepts, explain and analyze them briefly focusing on the relevant ones in light of theory development for this research project.

One of the main challenges in marketing research is a lack of common understanding and definition of key concepts. Particularly in the area of consumer behavior focusing on loyalty and satisfaction literature hardly shows common definitions, often the concepts overlap or even contradict each other. Important terms such as satisfaction or relationship are regarded differently in various research traditions. This may result from the fact that marketing research is influenced by traditions of psychology, sociology, behaviorism and economics. Thus, the identification of relevant latent variables is sometimes challenging. The following sections present a choice of theories and constructs for definition and measurement of those relevant concepts. The aim is a precise definition and overview with a focus on those serving as a theoretic basis for the research at hand, guiding model development, and composing parts of the actual research model.

Due to the nature of mobile services an integrative approach, focusing on service quality and diffusion theory seems promising for model development.

Apart from literature review for theory building the methodological framework needs consideration. As Structural Equation Modeling (SEM) or causal modeling as it is also referred to, seems to be the appropriate statistical method to test multiple indicator latent variable models. This method is also explained in the methodological building block.

Electronic service quality in general and mobile service quality in particular have specific requirements regarding theory building. Here

i) A lot of theory in service quality stems from a comparison of expected and actual service performance. This view is not supported in this research as customers often do not know what to expect when it comes to new technology products and the measurement instruments' complexity is unnecessarily increased.

ii) Traditional service quality literature is built on considerations of human to human interaction. With regard to mobile services this stream of research can serve as a guideline but one has to consider the fact that mobile services primarily involve human computer interaction.

iii) Adoption and diffusion theories bear a different sort of short coming not considering behavioral outcomes of system use. Additionally these theories often do not define "use" precisely enough. Testing a system once or trying a new service without repeating it is not the ultimate goal of service providers. In this research repeat use and behavioral consequences of the service use is of interest, thus, recursive models with the endogenous variable attitude towards the system/service use are unsatisfactory. Therefore, model development for this research project goes beyond traditional adoption models integrating those into the extensive body of knowledge from service quality and consumer behavior theory.

vi) Some of the dimensions in adoption and service quality models capture the general domain of IS adoption and service quality fairly well but they need to be reconsidered and evaluated (from an empirical view) if they are distinct enough for mobile service quality assessment and mobile consumer behavior.

The following sections, representing the theoretical building block for this project, present adoption and diffusion theories, the concept of service quality, related Internet service quality models, and, with respect to the empirical work involved in this research project, the underlying assumptions of causal modeling.

5.2 Diffusion and Adoption Theories

This chapter presents and discusses the most important theories concerning adoption with respect to mobile services and their importance for this research project. Uses and Gratifications research will not be discussed as it has generally been criticized for its limited theoretical foundation (Lin, 1996) and low explanatory power (LaRose, Mastro et al., 2001). The following table presents some models and theories of individual acceptance.

Table 6: Models and Theories of Individual Acceptance (See also Venkatesh, Morris, et al. 2003)

Model/Theory	Authors	Core Constructs
Innovation Diffusion Theory (IDT)	(Rogers, 1962)	Relative Advantage, Ease of Use, Image, Visibility, Compatibility, Results Demonstrability, Voluntariness of Use
Theory of Reasoned Action (TRA)	(Ajzen and Fishbein, 1980; Fishbein and Ajzen, 1975)	Attitude Toward Behavior, Subjective Norm
Technology Acceptance Model (TAM)	(Davis, 1989; Davis, Bagozzi et al., 1989; Venkatesh and Davis, 2000)	Perceived Usefulness, Perceived Ease of Use, Subjective Norm (TAM 2)
Motivational Model (MM)	(Davis, Bagozzi et al., 1992; Vallerand, 1997; Venkatesh and Speier, 1999)	Extrinsic Motivation, Intrinsic Motivation
Theory of Planned Behavior (TPB)	(Aijzen, 1991; Harrison, Mykytyn et al., 1997; Mathieson, 1991; Taylor and Todd, 1995b)	Attitude Toward Behavior, Subjective Norm, Perceived Behavioral Control
Task Technology Fit (TTF)	(Goodhue and Thompson, 1995)	Task Characteristics, Performance Impacts, Utilization, Technology Characteristics, Individual Characteristics
Unified Theory of Acceptance and Use of Technology (UTAUT)	(Venkatesh, Morris et al., 2003)	Performance Expectancy, Social Influence, Facilitating Conditions, Effort Expectancy, Behavioral Intention, Use Behavior, Attitude Toward Using Technology
Model of PC Utilization (MPCU)	(Thompson, Higgins et al., 1991), (Triandis, 1977)	Job Fit, Complexity, Long-term Consequences, Affect Towards Use, Social Factors, Facilitating Conditions
Social Cognitive Theory (SCT)	(Compeau and Higgins, 1995; Compeau, Higgins et al., 1999)	Outcome Expectations – Performance, Outcome Expectations – Personal, Self-efficacy, Affect, Anxiety

5.2.1 Diffusion of Innovations

The diffusion of an innovation traditionally has been defined as the process by which an innovation *"is communicated through certain channels over time among the members of a social system"*. (Rogers, 1983, 5)

The factor of communication channels, mentioned in the definition can be considered as one of the most important. The information about an innovation is communicated to or in a social system via mass media or interpersonal communication. The person's social system consists of different perceptions on media types and their reliability. Some might rely on mass media and others on interpersonal channels in their information seeking process. Some might rely on word-of- mouth and peer influence whereas others rely on the Internet or news papers for information.

Past research offers a solid foundation for theory development, one of those foundations is Rogers (1962) work on the diffusion of innovations. This work has stimulated various surveys and has been validated in various areas (Au and Enderwick, 2000).

Rogers (1962) defined five factors influencing consumer's adoption decisions, relative advantage, compatibility, complexity, communicability, and trialability. All of those are now explained in more detail.

Relative advantage: The degree to which an innovation is perceived as better than the idea it supersedes. The degree of relative advantage may be measured in economic terms, but social-prestige factors, convenience, and satisfaction are also important components. It also matters that there is some relative advantage of completing the transaction via an alternative medium. This was explored by various researchers investigating the advantage of transactions via the Web (Moore and Benbasat, 1991; Rogers, 1995; Seybold, 1998)

Complexity: The degree to which an innovation is perceived as difficult to understand and use is complexity. Some innovations are rapidly understood by most members of a social system; others are more complicated and will be adopted more slowly. New ideas that are simpler to understand will be adopted more rapidly than innovations that require the adopter to develop new skills and understandings.

Compatibility is the degree to which an innovation is perceived as being consistent with the existing values, past experiences, and needs of potential adopters. An idea that is incompatible with the values and norms of a social system will not be adopted as rapidly as an innovation that is compatible.

Trialability is the degree to which an innovation may be experimented with on a limited basis. New ideas that can be tried on the installment plan will generally be adopted more quickly than innovations that are not divisible.

Observability is the degree to which the results of an innovation are visible to others. Obviously, innovations and their success have a great deal to do with how they are perceived by the potential and actual consumers. The easier it is for individuals to see the results of an innovation, the more likely they are to adopt it.

Rogers (1995, p.16) puts forth that *"innovations that are perceived by individuals as having greater relative advantage, compatibility, trialability, observability, and less complexity will be adopted more rapidly than other innovations"*.

In order to be successful the innovation has to be communicated to or in the social system. Communication can be carried out via mass media or via personal communication. Rogers (1962) defines diffusion as an inherently social process. Thus, information can be transmitted via interpersonal channels, with a person adopting an innovation on the basis of the evaluation of an individual like them, or mass media channels, such as news paper, television, and radio. To analyze the influence of various media one can apply the Bass model for analysis (Bass, 1969). This model assumes that two types of communication channels influence potential adopters; namely, the mass media channel and interpersonal channels.

The social or cultural structure affects the diffusion and adoption of innovation in a system. A social system is defined as *"a set of interrelated units that are engaged in joint problem-solving to accomplish a common goal"* (Rogers, 1995, 23). The diffusion of an innovation within a social system, such as an organization, is influenced by norms of the system. Additionally, the opinion leaders within that social system will affect adoption as well as the change agent. Factors influencing the process are: the structure of the social system, norms, the role of opinion leaders and the consequence of the innovation (Rogers, 1995, 23).

Tomatzky and Klein (1982) found that three of Roger's innovation characteristics (perceived advantage, compatibility and complexity) were consistently related to adoption behavior. Davis derived his constructs perceived usefulness of the technology and perceived ease of use of the technology from the diffusion of innovation perspective (Rogers, 1995).

Diffusion models have been used to explain the adoption of mobile tourism services (Corigliano and Baggio, 2004) but only give insights on an aggregate level. An other application of diffusion models with regard to mobile commerce is the survey of Kleijnen et al. (2004) where perceived risk plays a critical role in the adoption process, followed by complexity and compatibility.

5.2.2 Social Cognitive Theory

The Compeau and Higgins' (1995) paper discusses the role of individual's beliefs about their abilities to use a computer. The model is based on Bandura's Social Cognitive Theory (1986). The survey investigates emotional reactions to computers (affect and anxiety), as well as outcome expectations and computer self-efficacy on computer use.

Self-efficacy and outcome expectations are positively influenced by encouragement of others and others' use of computers. Self-efficacy is important in an organizational setting when it comes to successful information system (IS) implementation. This provides implications for organizational support, training, and implementation.

In a later survey the second part of the model (not including encouragement by others, others' use and support) was tested (Compeau, Higgins et al., 1999).

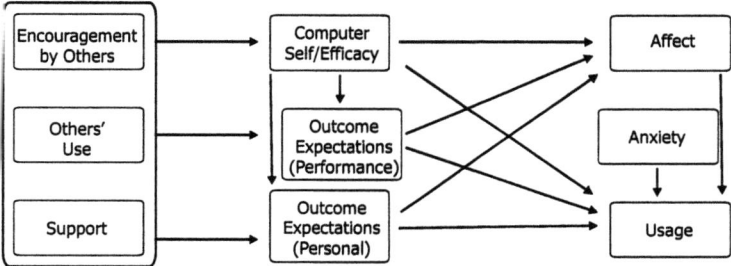

Figure 5: Compeau and Higgins SCT Model (1995)

In a later survey performance outcomes were found to influence affect and use with affect significantly related to use (Compeau, Higgins et al., 1999). Overall an individual's affective and behavioral reactions to information technology both impact self-efficacy an outcome expectations.

5.2.3 Theory of Planned Behavior & Theory of Reasoned Action

The Theory of Planned Behavior (TPB) is an extension of the Theory of Reasoned Action (TRA) and a well-established general theory of social psychology stating that specific beliefs influence given behavioral perceptions and resulting actual behavior (Aijzen, 1991; Ajzen and Fishbein, 1980). The building blocks of the TPB are salient beliefs, which are used to determine attitudes, social norm and behavioral control, consecutively determining intentions and behavior (see Figure 6).

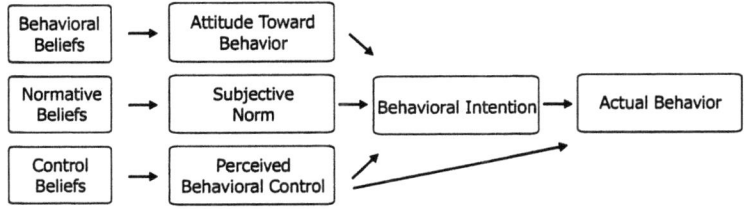

Figure 6: The Theory of Planned Behavior

Ajzen (2001) pointed out the ability of TPB to provide a useful theoretical framework for understanding and predicting the acceptance of new information technology. Empirical evidence suggests the explanatory power of TPB in the field of new information technology adoption. This includes a survey on the acceptance of telemedicine technology by physicians (Chau and Hu, 2002), the widespread adoption of virtual banking (Liao, Shao et al., 1999), executives adopting new information technology (Harrison, Mykytyn et al., 1997) and the acceptance of electronic brokerage services (Bahattacherjee, 2000).

5.2.4 The Technology Acceptance Model

A frequently cited adoption model in information systems is Davis' technology acceptance model (TAM) (Davis, 1989). The TAM adapted the Theory of Planned Behavior by incorporating technology in order to explain computer usage.

According to Davis et al. (1989, 985), the TAM goal is *"to provide an explanation of the determinants of computer acceptance that is general, capable of explaining user behavior across a broad range of end-user computing technologies and user populations, while at the same time being both parsimonious and theoretically justified."*

The TAM has been used to explain the adoption of telecommunication services such as telework (Hu, Chau et al., 1999), mobile telephones (Kwon and Chidambaram, 2000) and mobile commerce services (Pedersen, 2003). These studies suggest that the traditional TAM (Davis, 1989) needs modifications in underlying assumptions of usefulness when explaining the adoption of mobile services. They suggest the inclusion of social influence and behavioral control variables to explain the adoption process.

The TAM has successfully predicted and explained individual's intention to adopt and actual adoption across a variety of studies. Gefen and Straub (2000) provide a synopsis of TAM studies from 1989-2000. The model has been extended by including gender (Gefen and Straub, 1997; Venkatesh and Morris, 2000), age (Morris and Venkatesh,

2000), social norms (Pedersen and Herbjorn, 2003) and culture (Straub, Keil et al., 1997).

Pedersen et al. (2003) used an extended TAM to test the acceptance of a mobile parking service and found that with the extension (self expressiveness) the model satisfactory measured the adoption behavior.

The following figure depicts the TAM model.

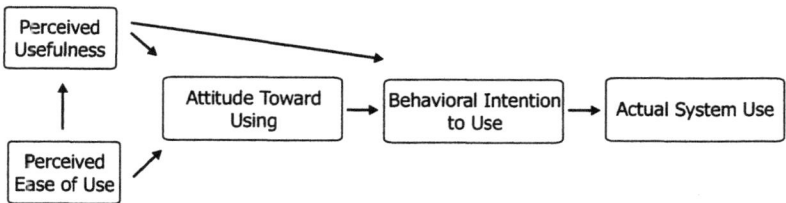

Figure 7: The Technology Acceptance Model

5.2.5 Task Technology Fit Model

User evaluations are attitudes or beliefs about something and have been used to measure different *"things"*. Goodhue's (1995) approach for developing the Task Technology Fit Model (TTF) was that IS literature acked a specific user evaluation construct with a theoretical perspective that links underlying systems to their relevant impacts. Some authors (Goodhue and Thompson, 1995; Jarvenpaa, 1989; Zirgus and B.K., 1998) integrated concepts from IS and organizational research. They discuss outcomes of matching group support applications with group task requirements on performance and process quality. In the ultimate TTF model Goodhue and Thompson (1995) combine utilization and fit as a technology must be utilized and must be of good fit with the tasks it supports. This concept is shown graphically in the following figure. The results of their study highlight the importance of the fit between technologies and users' tasks in achieving individual performance impacts.

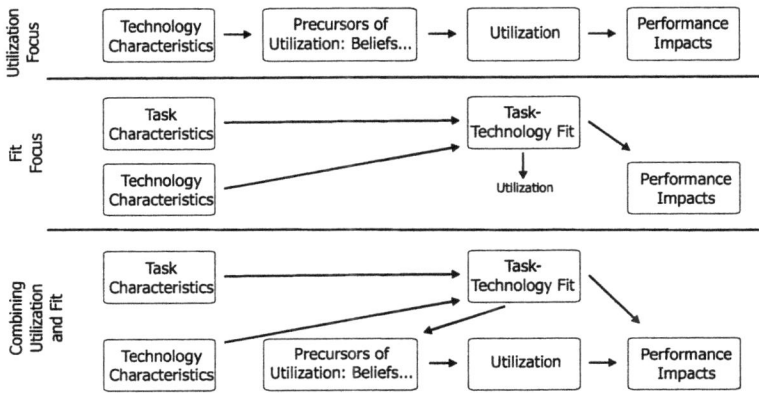

Figure 8: Task Technology Fit Model (Goodhue and Thompson, 1995)

With regard to the m-services industry TTF was applied to investigate the tasks performed via a variety of interaction devices such as wireless phones and PDAs (Wells, Sarker et al., 2002).

5.2.6 Motivational Model

Research from the field of psychology supported the general motivation theory as predictor for behavior. The core constructs are extrinsic motivation and intrinsic motivation. Extrinsic motivation could be an improved job performance, pay or promotions whereas intrinsic motivation would be that the user performs the activity per se because he/she wants to (Davis, Bagozzi et al., 1992). This theory has been adapted for the IT context in various studies (Davis, Bagozzi et al., 1992; Venkatesh and Speier, 1999).

5.2.7 Unified Theory of Acceptance and Use of Technology

The most recent model development and assessment was carried out by Venkatesh, Morris et al. (2003), leading technology acceptance researchers, with an attempt to evaluate existing models to build a unified model of technology acceptance.

The model was formulated based on conceptual and empirical similarities across eight competing technology acceptance models. These are:

1. Davis' Technology Acceptance Model (Davis, 1989; Davis, Bagozzi et al., 1989)
2. Roger's Innovation Diffusion Theory (Rogers, 1995)
3. The Theory of Reasoned Action (Fishbein and Ajzen, 1975)

4. The Motivation Model (Davis, Bagozzi et al., 1992)
5. The Theory of Planned Behavior (Aijzen, 1991)
6. The Combined Technology Acceptance and Theory of Planned Behavior (Taylor and Todd, 1995b)
7. The Model of PC Utilization (Thompson, Higgins et al., 1991; Triandis, 1977)
8. The Social Cognitive Theory (Compeau and Higgins, 1995; Compeau, Higgins et al., 1999)

The unified theory of acceptance and use of technology (UTAUT) comprises four core factors determining intention to use – performance expectancy, effort expectancy, social influence and facilitating conditions (Venkatesh, Morris et al., 2003). Gender, age, experience, and voluntariness of use moderate the key relationships in the model (Venkatesh, Morris et al., 2003).

The performance expectancy is according to Venkatesh Morris et al. (2003) the degree to which an individual believes that using a system will help to better attain rewards. In previous surveys performance expectancy (usefulness) has consistently been a strong predictor of intention (Davis, Bagozzi et al., 1989; Taylor and Todd, 1995b; Venkatesh and Davis, 2000).

Effort expectancy is the degree of ease associated with the use of systems. Three constructs from existing models capture the concept of effort expectancy: perceived ease of use (TAM), complexity (MPCU), and ease of use (IDT).

Social influence is the degree to which a person perceives that important others believe he/she should use a new technology. Subjective norm, image, and social factors were included in previous models to represent this factor.

Facilitating conditions the degree to which an individual believes that an organizational and technical infrastructure exists to support system usage. Perceived behavioral control (TPB), facilitating conditions (MPCU) and compatibility (IDT) capture this idea in previous research.

Consistent with all previously mentioned behavioral models intention has a significant positive influence on technology usage. UTAUT also includes the moderating variables gender, age, experience, and voluntariness of use.

Now, after describing the components of the model extensively, Figure 9 illustrates the UTAUT model.

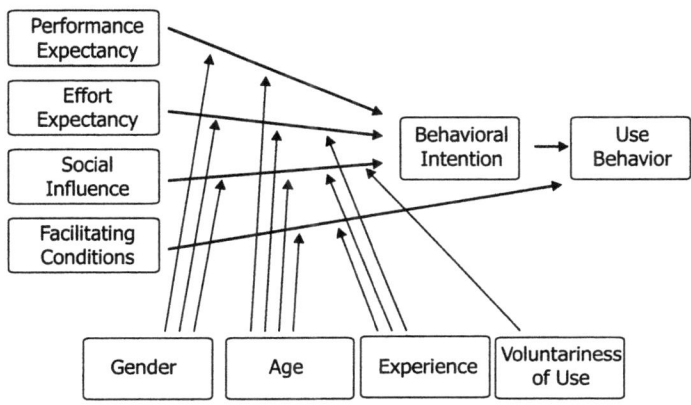

Figure 9: UTAUT Model (Venkatesh, Morris et al., 2003)

5.2.8 Comparative Analysis of Adoption Models

Venkatesh, Morris et al. (2003) provide an overview of model comarisons. The main findings are condensed here. Mathieson (1991) stresses that both TAM and TPB predict intention to use quite well with TAM easier to apply and TPB providing more specific information that can better guide development.

Taylor and Todd (1995b) conducted a survey testing the predictive quality of TAM and TPB. In their survey they found that a decomposed model consisting of TAM and TPB items provided a fuller understanding of behavioral intention.

While there have been some tests of models in an organizational setting, Plouffe, Hulland et al. (2001) are the only ones comparing models in that setting. All other comparison studies have been surveys among students.

The most extensive model comparison so far, carried out by Venkatesh, Morris et al. (2003) first tests eight models, among those are Theory of Reasoned Action, Technology Acceptance Model, Motivational Model, Theory of Planned Behavior, Combined TAM and TPB, Model of PC Utilization, Innovation Diffusion Theory and Social Cognitive Theory.

In the Venkatesh, Morris et al. (2003) survey all models are compared on all participants, which is a major advantage compared to previous studies. Most model comparisons were conducted in voluntary usage contexts. Thus, one has to be cautious generalizing these findings to mandatory usage scenarios. The Venkatesh et al. (2003) survey acknowledges this fact by examining both, voluntary and mandatory implementation contexts.

Table 7: Review of Model Comparisons (See also Venkatesh et al. 2003)

Model Comparisons	Models	Findings
Mathieson (1991)	TAM, TPB	The variance in intention explained by TAM 70%, TPB 62%.
Davis et al. (1989)	TRA, TAM	The variance in intention and use explained by TRA was 32% and 26%, and TAM was 47% and 51% respectively.
Taylor and Todd (1995b)	TAM, TPB, Decomposed TPB	The variance in intention explained by TAM was 52%, TPB was 57% and DTPB was 60%.
Venkatesh et al. (2003)	TRA, SCT, TAM/TAM2 TPB/DTPB, MM, UTAUT, C-TAM/TPB, MPCU, IDT	UTAUT outperformed the other eight models which explained between 17% and 53% of variance in user intentions to use information technology while UTAUT was confirmed with 70% of explained variance in user intentions.
Plouffe et al. (2001)	TAM, IDT	The variance in intention explained by TAM was 33% and IDT was 45%.

5.3 Service Quality

Service quality and satisfaction are two of the most complex constructs in marketing theory, which is reflected by a massive amount of publications on this subject. Particularly the measurement of service quality has attracted the attention of academics and practitioners (Roest and Pieters, 1995). The definition of service quality is a complex and difficult task and marketing literature offers myriad options.

Service quality resulting from the use of mobile services is of interest for this book. Furthermore, the behavioral outcomes resulting from service quality are explored. Quality has been focused on products and the technical perspectives of them resulting in Total Quality Management in the 90s. In this research the consumer's judgment of a service is the center of interest. According to Parasuraman, Zeithaml and Berry a consumer oriented definition is perceived quality as: "...*the consumer's judgment about an entity's overall excellence or superiority...*" Zeithaml in: (Parasuraman, Zeithaml et al., 1985, 43)

The following table helps identify the distinctions of the various constructs and highlight some differences.

Table 8: Dimensions of Perceived Service Quality and Related Constructs (Roest and Pieters, 1997)

Dimensions	Perceived Quality	Perceived Value	Customer Satisfaction	Product Attitude
Basis	Get	Give and get	Give and get	Give and get
Object	Product	Product	Consumer	Consumer to product
Content	Cognitive	Cognitive	Cognitive and affective	Cognitive, affective and conative
Context	Relative	Relative	Relative	Absolute
Aggregation	Transaction and relationship	Transaction and relationship	Mainly transaction	Mainly transaction

The following subchapters introduce the most important concepts of service quality and highlight the different assumptions on causal relationships between the constructs. The presentation of the models follows a chronological order as some build on findings of others.

5.3.1 Donabedian's Model

One of the eldest models dealing with service quality is Donabeidian's (1987) model. Coming from the health care industry Donabedian identifies three dimensions for the quality of medical services:
1. Structure
2. Process
3. Outcome (Donabedian, 1987; Donabedian, 1991)

Structure refers to the capabilities necessary to make the service. This is the technical and organizational ability of the service provider and the potential to use regarding the consumer.

Process comprises all activities involved in the completion of the service.

The result of the process is referred to as *Outcome*. In the health care industry this would be an improved condition of the patient.

Donabedian suggests linearity and sequential steps among these three components. The main achievement of Donabedian's view is still valid; service quality does not lie in the outcome but also in the process of completing the service. This model serves as a basis for subsequent attempts to model service quality.

In his more recent book on quality assurance Donabedian (2003) breaks quality improvement into two component parts: *systems design and resources* and *performance monitoring and adjustment*. These two components feed into each other.

5.3.2 Grönroos' Model

Grönroos' (1978; 1982; 1984) work on the deconstruction of service quality concludes that service quality's principal components are technical quality, functional quality and corporate image. He applies a target/actual comparison of expectation with the service in the way he did with the Confirmation/Disconfirmation-Paradigm in satisfaction theory. The following figure illustrates how technical quality and functional quality are included from a customer's point of view.

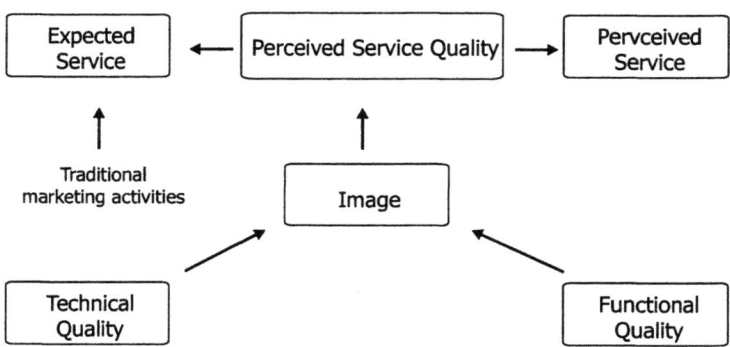

Figure 10: Grönroos' (1984) Service Quality Model

The technical quality dimension focuses on the *what* while the functional quality dimension focuses on the *how*. *What* the customer receives is a result of his interactions with a firm. This can be measured by the customer objectively. The *how* of receiving the service is also important to the customers but they cannot measure it as objectively. The technical quality includes know how, technical solutions, machines, and computerized systems. The functional quality includes internal relations, behavior, service mindedness, appearance, accessibility, customer contacts and attitudes. Grönroos is explicit with respect to the contrasting *hard* and *soft* aspects of the service output.

All partial evaluations are influenced by the existing image of a service provider. Corporate and/or local image is particularly important for most services (Grönroos, 1988). It has an impact on the perception of quality and can even be considered as a filter. Finally, good perceived quality is achieved when the experienced quality meets the expectations of the customer. If expectations are not realistic the perceived quality will be low no matter if the quality – objectively measured – was good (Grönroos, 1988). Grönroos defines six criteria of good perceived service quality that can be accounted to the dimensions:
- Professionalism and skills
- Attitudes and behavior

- Accessibility and flexibility
- Reliability and trustworthiness
- Recovery
- Reputation and credibility

Marketing activities include advertising, public relations and sales campaigns that can be controlled by the firm. In contrast word-of-mouth and image factors are only indirectly controlled by the firm.

Later, considering developments in the service quality research with respect to the constructs *satisfaction* and *service quality*, Grönroos suggests to rename his model *"Perceived Service Features"* to stress the focus on the attributes of a service (Grönroos, 2001). Some authors criticize the model due to insufficient empirical tests and neglecting the external factor. However, it gives valuable insights into the quality dimensions.

5.3.3 Parasuraman, Zeithaml, Berry

The model developed by Parasuraman, Zeithaml and Berry (1985) was based on an explorative study of experts' evaluations and focus groups in four service businesses. The model's basic idea is the existence of gaps between customers' expectations and perceived performance. Some findings were industry specific but the majority were common across industries and, thus, encourage the formulation of a general model. This was the basis for the subsequent development of a measurement instrument called SERVQUAL (Parasuraman, Berry et al., 1991; Parasuraman, Zeithaml et al., 1988). The model is described in Figure 11.

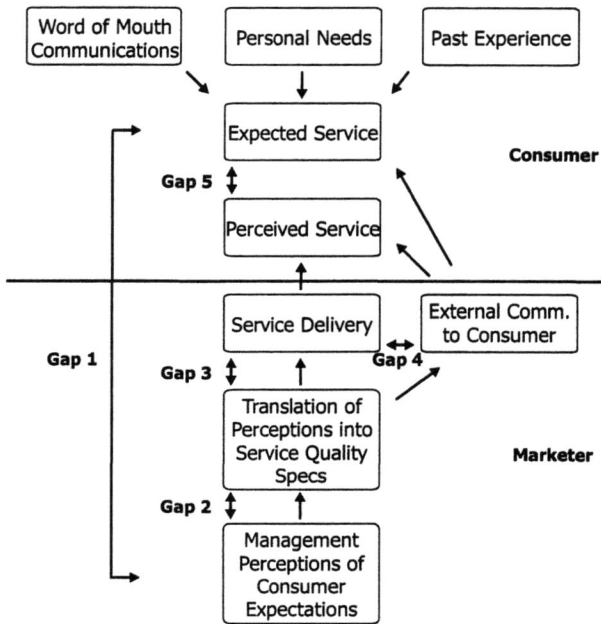

Figure 11: Service Quality Model of Parasuraman/Zeithaml/Berry (1985)

Gap 1: Consumer expectation – management perception gap. This gap implies a difference between customer expectations and executive perceptions about them. Discrepancies between those two exist.

Gap 2: Management perception – service quality specification gap. This is the discrepancy of management perceptions of consumer expectation and the firm's service quality specifications that influences the customers' service quality perceptions.

Gap 3: Service quality specifications – service delivery gap. Even with the existence of guidelines performing services may not be a certainty. A firm's employees have a strong influence on the service quality perceived by the customers. This is the gap between service quality specifications and actual service delivery.

Gap 4: Service delivery – external communications gap. This gap occurs when a company promises more in its marketing communications than the services can deliver in reality.

Gap 5: Expected service – perceived service gap. The explorative results showed that the key to good service quality is meeting or exceeding what consumers expect from a service. The gap is the extent to which the expected service is not met by the perceived service (Parasuraman, Zeithaml et al., 1985).

From a customer's perspective gap five is the most important one. For companies gaps 1-4 are most important and their direction (positive or negative) has an impact on gap 5.

In a later survey they stress the relationship between service quality and customers' intentions concerning information systems (Berry and Parasuraman, 1997). These statements of intentions comprise *"Loyalty to Company"*, *"Propensity to Switch"*, *"Willingness to Pay More"*, *"External Response to Problem"* and *"Internal Response to Problem"*. The information quality measures if the information is relevant, precise useful in context, credible, understandable, and timely.

5.3.4 Measurement of Service Quality

5.3.4.1 SERVQUAL

There are several measurement instruments for service quality. This chapter presents SERVQUAL measuring GAP number five from the previously explained GAP model (Parasuraman, Zeithaml et al., 1988).

Parasuraman, Zeithaml and Berry (1988) described a model of five dimensions of service quality: tangibles, reliability, responsiveness, assurance, and empathy.

- Tangibles: Tangibles are the physical facilities, equipment and appearance of personnel for instance.
- Reliability: Reliability refers to the ability to perform the desired service dependably, accurately, and consistently.
- Responsiveness: Responsiveness is the willingness to provide prompt service and support customers.
- Assurance: This concept includes employees' knowledge, courtesy, and the ability to convey trust and confidence.
- Empathy: Provision of caring and individualized attention to consumers is understood as empathy. (Berry, Parasuraman et al., 1988; Parasuraman, Berry et al., 1991; Parasuraman, Zeithaml et al., 1988).

Although the model was originally constructed for the traditional service sector it can be adapted to the new technology sector. In this setting tangibles would be the user interface, reliability can be the error probability of the system itself e.g. an unstable phone connection would influence the reliability negatively. Responsiveness can relate to the fact that mobile services should be highly interactive but yet, at the same time, should be unobtrusive. Assurance is another aspect named by Berry, Parasuraman and Zeithaml (1988), which can translate into the extent to which the service provides trustworthy information and the user can have confidence in the services. Finally, empathy with regard to mobile services refers to the degree of personalization of the service provided, which is not yet fully applicable in this setting as services are yet predominantly pulled not pushed. Once push services increase the degree of individualism and personalization will become even more important.

SERVQUAL consists of 22 items for customers' expectations and perceptions of service quality, using a seven point Likert scale for measurement.

Criticism regarding SERVQUAL falls mainly into two areas: i) the rather vague formulation of the items and ii) asking the consumer about expectations.

The first criticism involved the fact that the SERVQUAL items are formulated in a general manner to allow application in as many industries as possible. Thus, in some cases the items are too generic and need to be adapted to the specific setting. As a result the advantage of SERVQUAL, being standardized for an application in various areas often is not applicable (Ostrowski, O'Brien et al., 1993).

Secondly, measuring expectations has been criticized as in practice expectations are often mixed with performance evaluation when not asked prior to consuming the service. Quality is the subjective perception of an individual customer and often is not fulfilled. In fact, the customer can perceive high quality even if the (admittedly) unrealistic expectations were not quite met.

5.3.4.2 SERVPERF

Cronin et al. (1992; 1994) criticise SERVQUAL and also provide an alternative instrument called SERVPERF. Other authors also support the argument that simple performance based measures of service quality are satisfying and outperforming the expectations-performance approach (Bolton and Drew, 1991; Woodruff, Cadotte et al., 1983).

Their reasoning aligns with defining service quality as an attitude developed over time and that satisfaction is a transaction-specific measure. Oliver (1980) for instance suggests that attitude is initially a function of expectations and subsequently a function of the prior attitude toward the present level of satisfaction.

Reflecting on criticism on SERVQUAL Parasuraman, Zeithaml and Berry (1994a; 1994b) suggest a three column format which generates separate ratings of desired, adequate, and perceived service with three identical side-by-side scales. This format is longer, though, and might lead to respondents' resistance to fully complete the questionnaire.

With regard to the development of the SERVQUAL instrument Cronin and Taylor (1992) raise concerns related to the empirical evidence which suggests that the five components are not consistent. However, the validity of the 22 items appears to be well supported by the procedures used to develop them and their subsequent use as reported in literature. Therefore, Cronin and Taylor (1992) stick with the basic 22 items. In summary they found that the only performance oriented SERVPERF measure explains more variance than the SERVQUAL instrument.

Parasuraman et al. (1994b) comparie of four alternative measures including SERVQUAL and SERVPERF. They conclude that both, SERVPERF and a summary disconfirmation measure outperform SERVQUAL. However, they recommend the continued use of their measure which is quite surprising.

5.3.5 Internet Quality

There are research attempts that focus on the development of a model for Web site quality, also known as WebQual. The next section introduces such models, two of them named WebQual one developed by Barnes and Vidgen (2002), the other one by Loiacono, Watson and Goodhue (2002). One called Sitequal by Yoo and Donthu (2001), eTailQ by Wolfinbarger and Gilly (2003), EC-SERVQUAL by Wang and Tang (2003), DeLone and McLean's (1992) IS success, E-S-Qual (Parasuraman, Zeithaml et al., 2005), and attitude towards the site (Chen, Clifford et al., 2002).

The models discussed earlier focus on human interaction. This setting is not fully applicable as mobile phones are electronic devices. Thus, drawing on concepts developed for the Internet may yield insights for the mobile medium and support model development in course of this research project.

5.3.5.1 E-S-Quality

Recent (Parasuraman, Zeithaml et al., 2005) studies looked into what constitutes e-service quality and Zeithaml, Parasuraman and Malhotra (2000) state *"e-service quality (e-SQ) is the extent to which a website facilitates efficient and effective shopping, purchasing, and delivery."*

Researchers agree that SERVQUAL is a good predictor of overall service quality and adequate for measuring IS service quality (Kettinger and Lee, 1994; Pitt, Watson et al., 1995; Watson, Pitt et al., 1998).

The process of developing e-SQ involved six steps i) a literature and qualitative study ii) a preliminary scale containing 121 items representing 11 e-service quality dimensions iii) an online survey iv) development of a parsimonious scale through an iterative process v) a confirmatory factor analysis and validity test on the final scale vi) a reconfirmation of reliability and validity (Parasuraman, Zeithaml et al., 2005). The following table explains the dimensions included in the model and what these dimensions are anteceding.

Table 9: Dimensions of e-SQ (Adapted from Zeithaml (2000), Parasuraman (2005))

Dimension	Explanation	Anteceding
Reliability	Involves the correct functioning of the site and the accuracy of service promises, billing, and product information.	Perceived Control
Responsiveness	Means quick response and the ability to get help if there is a problem or question.	Perceived e-Service Quality
Access	Is the ability to get on the site quickly and to reach the company when needed.	Perceived Convenience
Flexibility	Involves choice of ways to pay, ship, buy, search for, and return items.	Perceived Convenience/Control
Ease of navigation	Means that a site contains functions that help customers find what they need easily, possesses a good search engine, and allows the customer to move easily and quickly through the pages.	Perceived Convenience
Efficiency	Means that a site is simple to use, structured properly, and requires a minimum of information to be input by the consumer.	Perceived Convenience
Assurance/Trust	Involves the confidence the customer feels in dealing with the site and is due to reputation of site and products/services as well as clear and truthful information.	Perceived e-Service Quality
Security/Privacy	Involves the degree to which the customer believes the site is safe from intrusion and personal information is protected.	Perceived Control
Price Knowledge	Is the extent to which the customer can determine shipping price, total price, and comparative prices during the shopping process.	Perceived e-Service Quality
Site Aesthetics	Relates to the appearance of the site.	Perceived e-Service Quality
Customization/ Personalization	Is how much and how easily the site can be tailored to individual customer's preferences, histories, and ways of shopping.	Perceived Control

The table shows factors anteceding perceived control and perceived convenience, which in turn are antecedents of perceived e-service quality. Perceived price antecedes perceived value, which is influenced by the e-service quality. The whole causal chain ends in purchase/repurchase. Both, experienced and less experienced respondents mentioned similar attributes to evaluate e-SQ, thus, according to the authors (Zeithaml, Parasuraman et al., 2000) it should be possible to develop a general scale. Expectations play an important role in perceived service quality. However, participants of the e-SQ expressed that it is hard for them to articulate their e-SQ expectations.

In a later publication Parasuraman Zeithaml and Malhotra (2005) again worked on a multiple-item scale for measuring service quality delivered by Web sites for online shopping. In this research they stress the importance of value as an additional overall assessment such as e-SQ, which in turn influence behavioral intentions and actual behavior.

Their final e-SQ scale includes efficiency, fulfillment, system availability and privacy. The second outcome was an e-recovery service quality scale (e-RecS-Qual) including responsiveness, compensation and contact (Parasuraman, Zeithaml et al., 2005).

5.3.5.2 EC-SERVQUAL

Wang and Tang (2003) used the SERVQUAL model as starting point and developed an EC-SERVQUAL model containing four dimensions: reliability, responsiveness, assurance and empathy. This study adopted the two column format of SERVQUAL to develop the EC-SERVQUAL.

The measurement instrument was modified to apply to the e-commerce context. The instrument was originally developed for a human to human interaction setting, thus, questions with regard to a sales person had to be rephrased to ask about the screen design etc. After adaptations each item was turned into an expectation and perception measurement. The final instrument with good reliability and validity provides researchers with 16 measurements for explaining differences across results (Wang and Tang, 2003).

5.3.5.3 WebQual by Barnes and Vidgen

Barnes and Vidgen (2001a) developed a web quality assessment instrument, improved and changed it based on newer findings. The instrument, based on Bossert's (1991) theory of quality function deployment, can be found in four versions, WebQual 1.0 to WebQual 4.0.

WebQual version 2.0 included 10 dimensions (aesthetics, understanding the individual, communication, access, security, credibility, navigation, competence, responsiveness, and reliability). An analysis of the WebQual 3.0 data lead to three dimensions of e-commerce Web site quality: information quality, service interaction quality, and usability. Their latest WebQual version 4.0 (Barnes and Vidgen, 2003) is reduced to a quite parsimonious model of four dimensions: usability, design, information, and service. Site quality has been replaced by usability as it emphasizes the user and the users' perceptions rather than the designer. The usability dimension draws on literature from the field of human computer interaction (Davis, 1989). Usability is concerned with users' interaction and perception of a Web site.

Applications of WebQual include UK business school websites (Barnes and Vidgen, 2000), Internet bookshops (Barnes and Vidgen, 2001d), small companies (Barnes and Vidgen, 2001b), and online auction houses (Barnes and Vidgen, 2001c).

5.3.5.4 WebQual by Loiacono, Watson and Goodhue

As opposed to their colleagues Loiacono, Watson and Goodhue (2000) focus on TAM and TRA as starting points to develop their measure for Web site quality. According to the authors those theories provide a strong conceptual basis for a link between beliefs and behavior that can be applied in the web context.

The authors criticize Barnes and Vidgen's model for a lack of content validity and a too small sample size in the first three versions inapt to provide reliable factor scores.

Loiacono et al.'s (2000; 2002; 2000) proposed WebQual instrument measures 12 core dimensions with 4 umbrella terms: usefulness (informational fit-to-task, tailored communications, trust, response time), ease of use (ease of understanding, intuitive operations), entertainment (visual appeal, innovativeness, emotional appeal) and finally, complimentary relationship (consistent image, on-line completeness, relative advantage). The application of the Loiacono et al. (2002) WebQual instrument in combination with Hofstede's cultural dimensions showed that masculinity and long-term orientation are associated with higher Web site quality expectations (Tsikriktsis, 2002). Another application of this instrument showed that only three dimensions, informational fit-to-task, transaction capability and response time, were significant predictors of shopper satisfaction (Kim and Stoel, 2004).

5.3.5.5 DeLone and McLean's Model of IS Success

DeLone and McLean (1992) conducted an extensive survey on the success factors of information systems. According to them, IS success can be represented by the quality characteristics of the IS itself. These are the system quality, the quality of the output of the IS, and the information quality. The consumption or use of the IS relates to the users' response to or satisfaction with the IS. The effect of the system on the behavior of the user is the individual impact and the effect on the organizational performance is the organizational impact.

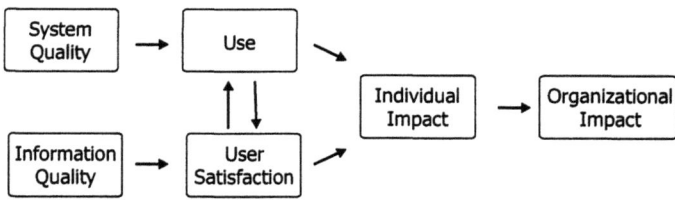

Figure 12: DeLone and McLean's (1992) Model of IS Success

5.3.5.6 Sitequal

The items of Sitequal were generated mainly on consumers descriptions as it focuses on individual perceptions of quality (Yoo and Donthu, 2001). The aim was to develop a reliable, valid, multidimensional, and parsimonious instrument.

In the first phase of developing the measurement instrument 92 site characteristics were identified and trimmed to 54 descriptions after excluding redundancies. Then the 54 characteristics were included in a questionnaire with 50 items remaining after the first analysis. Further model re-specification and refinement of the measurement instrument lead to four factors made up of nine items. These are ease of use, aesthetic design, processing speed, and security.

5.3.5.7 eTailQ

Wolfinbarger and Gilly (2003) developed eTailQ, a model explaining and predicting quality of online shopping. They argue that online consumers are aware of their need for privacy and security and above that writers and scholars have stressed the unique capabilities of the online medium to provide interactivity, personalized experiences, community, content, increased product selection, and information. This suggests that traditional concepts of service and retailing quality may be inadequate in an online context (Wolfinbarger and Gilly, 2003).

It is of interest what attributes of quality, satisfaction, and loyalty are important to online users. The measurement scale was developed based on online and offline focus groups, a sorting task, and an online survey of a customer panel with the aim to go beyond a sole website assessment to an e-tail service quality instrument. Figure 13: eTailQ shows the model with the final four factors extracted.

The four factors are:
- Fulfillment/reliability: it is the accurate display of a product and the customers receive what they expect in the time frame promised.
- Website design: it includes all elements of a consumer's experience on the Web site.
- Customer service: refers to a willing service provider that responds to the customer quickly.
- Security/privacy: takes credit cad payments and privacy of shared information into account.

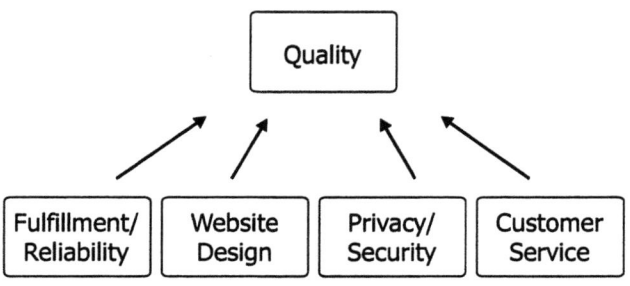

Figure 13: eTailQ (Wolfinbarger and Gilly, 2003)

5.3.5.8 Attitude Toward the Site – A_{ST}

In the beginning of the Internet, marketers thought the Internet would be quite measurable (Gibson, 1997) but soon realized that this in not easy at all. Previous research on Attitude toward the Ad (A_{AD}) has shown that the attitude is an indicator of advertising effectiveness. Haley and Baldinger (1991) found that how well viewers liked an ad was the best predictor of sales.

Analogous to that Chen and Wells (1999) assume that attitude toward the site is an equally useful indicator of site value. After collecting items from related surveys and conducting a quantitative survey the authors found that three factors influenced the attitude toward the site. These factors were: entertainment, informativeness, and organization (Chen, Clifford et al., 2002). The following figure shows the proposed model.

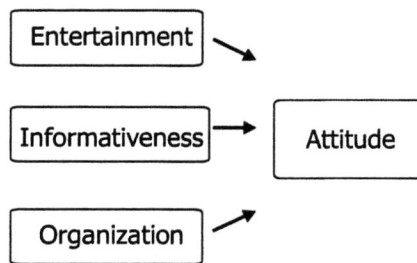

Figure 14: Attitude Toward the Site

5.4 Behavioral Consequences of Perceived Service Quality

Since most service quality and adoption models end with service quality or technology use further relevant constructs and theories are explored. The concept of loyalty is important since this research project does not focus on trial or one time usage but on repeat usage. Furthermore, value is explored since often perceived value is the main driver of repeat usage.

5.4.1 Loyalty

Loyalty is a construct in social science and the term is often used in everyday life. Jacoby and Chestnut (1978) cited 53 definitions in their review on loyalty. Yet, there is no consent on a single definition. Oliver, one of the most cited scientists in loyalty research defines loyalty as a

"...deeply held commitment to rebuy or repatronize a preferred product or service consistently in the future, despite situational influences and marketing efforts having the potential to cause switching behavior. (Oliver, 1997, 392)

The modeling of loyalty has a long history in marketing literature dating back to the early 1920s (Copeland, 1923). The majority of early studies conceptualized loyalty behaviorally. Some authors focused on the sequence in which brands were purchased; others measured loyalty through the proportion of purchases of a specific brand. A third group focused on stochastic measures like probability of purchase. Some combined behavioral criteria for empirical studies.

To give insights into loyalty research the following chapters present the important loyalty concepts.

5.4.1.1 Day's Two-Dimensional Loyalty Concept

Day (1969) viewed brand loyalty as repeated purchases prompted by a strong internal disposition. He proposed loyalty indexes based on attitudinal and behavioral measures and criticized the use of only behavior based loyalty measures. The main idea is that these do not distinguish between true loyalty and spurious loyalty. He argues that:

"The key point is that these spuriously loyal buyers lack any attachment to brand attributes and they can be immediately captured by another brand that offers a better deal..." (Day, 1969, 30)

Thus, he adds a two-dimensional conceptualization of loyalty adding an attitudinal dimension to the behavioral component.

5.4.1.2 Loyalty According to Jacoby, Chestnut and Kyrner

Jacoby and Chestnut (1978) criticized the behavioral measures as lacking conceptional basis and capturing only the static outcome of a dynamic process. In line with Day, Jacoby and Kyrner (1973, 2) also propose a two dimensional loyalty definition

- *"...brand loyalty is*
- *the biased (i.e. nonrandom)*
- *behavioral response*
- *expressed over time*
- *by some decision making unit,*
- *with respect to one or more alternative brands, and*
- *is a function of psychological (decision-making, evaluative) processes..."*

Jacoby and Chestnut's analysis concludes that consistent purchasing as indicator of loyalty could be invalid.

This research project supports the reasoning to not only consider one factor, e.g. buyer behavior or a positive attitude, as the determinant of loyalty. In line with this reasoning it is proposed to include two components in the loyalty construct, thus, this survey will use a two dimensional loyalty construct.

5.4.1.3 The Dick and Basu Loyalty Approach

As mentioned earlier Day originated the two dimensional loyalty definition. In addition to this two dimensionality, Day distinguished true/intentional loyalty and spurious loyalty. Purchases guided by situational aspects such as special offers reflect spurious loyalty. The traditional definitions do not attempt to understand the factors underlying repeat purchase. High repeat purchase may reflect situational factors whereas low repeat purchase may simply indicate different usage situations like variety seeking, or lack of brand preferences. Thus, behavioral definitions are insufficient to explain brand loyalty.

As a consequence Dick and Basu (1994) developed the conceptual framework of customer loyalty presented in Figure 15.

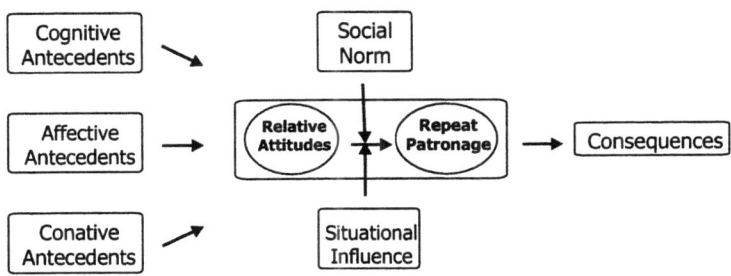

Figure 15: A Framework of Customer Loyalty (Dick and Basu, 1994)

The model includes three types of antecedents of relative attitudes (cognitive antecedents, affective antecedents and conative antecedents). Attitude represents an association between an object and an evaluation. The attitude towards a brand could be positive but still might not result in repeat purchase over time because of comparable or greater attitudinal extremity toward other brands. Thus, relative attitudes provide a stronger indication of repeat patronage. Cross-classifying relative attitudes and repeat patronage leads to four loyalty conditions: no loyalty, spurious loyalty, latent loyalty, and loyalty. The classification is shown in Figure 16.

		Repeat Patronage	
		High	Low
Relative Attitude	High	Loyalty	Latent Loyalty
	Low	Spurious Loyalty	No Loyalty

Figure 16: Types of Loyalty according to Dick and Basu (1994, 101)

Now the types of loyalty as shown in above figure are explained in further detail.

No loyalty. Low relative attitude combined with low repeat patronage is leading to the absence of loyalty. First, this could occur when a new product is introduced and/or in the case of inability to communicate distinct product advantages. Second, low relative attitude may be due to specific market place dynamics where competing brands are seen as similar. A marketing manager in this case can make efforts to create spurious loyalty through manipulation of situational exigencies or social norms. This is achieved through favorable locations for brands or aggressive trade promotions (Dick and Basu, 1994).

Spurious loyalty. Low relative attitude, with high repeat patronage is spurious loyalty. This type is characterized by non-attitudinal influences on behavior such as subjective norms and situational effects. This is similar to inertia, in that a customer perceives little difference between the brands of a low involvement category and, thus, does not change brands. A strengthening of the value of social bondage may also lead to repeat orders (Dick and Basu, 1994).

Latent loyalty. A high relative attitude accompanied with low repeat patronage reflects latent loyalty. This is a concern for marketers in a marketplace environment with subjective norms and situational effects equally or even more influential than attitudes in determining patronage behavior. A person for example may have a high relative attitude toward a restaurant but still prefer going to a different one due to varying preferences of meal companions. It appears unnecessary to increase an existing and already high positive attitude, thus, marketing goals should be focused on addressing the normative/situational constraints.

Loyalty. Loyalty, the desired of the four conditions, signifies a favorable correspondence between relative attitude and repeat patronage. The discussion on relative attitude suggests that, provided customers perceive differences among competing brands, the level of attitude does not have to be the highest. However, it is desirable. The competition will try to induce spurious loyalty through manipulation of situational factors, decrease perceived differentiation with the leading brand reducing its relative attitude (Me-too strategy) or through an increase of perceived differentiation in its favor through competitive claims of superiority (Dick and Basu, 1994).

The concept of Dick and Basu was often used and frequently empirically tested (Oliver, 1997; Oliver, 1999; Pitchard, Howard et al., 1992).

5.4.1.4 Oliver's Dynamic Loyalty Perspective

Oliver's work is based on Jacoby and Kyner (1973) and Dick and Basu (1994) adding a dynamic perspective. According to Oliver there is an ultimate loyalty where the customer will pursue his request against all odds and at all cost. Proactive loyalty occurs when a customer frequently rebuys a product and will not change to an other (Oliver, 1997). In contrast to that, situational loyalty appears in certain situations with external influence.

The basic idea of Oliver's model is that customers become loyal in a cognitive way first, then later in an affective, still later in a conative manner and finally in a behavioral manner. This is described as the *action inertia* (Oliver, 1999). At each stage the commitment and emotional connection increases.

The four-stage loyalty model has different vulnerabilities, depending on the stage and the nature of the customers' commitment. This is summarized in Table 10.

Table 10: Loyalty Phases with Corresponding Vulnerabilities (Oliver, 1999)

Stage	Identifying Marker	Vulnerabilities
Cognitive	Loyalty to information such as price, features, and so forth.	Actual or imagined better competitive features or price through communication (e.g. advertising) and vicarious or personal experience Deterioration in brand features or price. Variety seeking and voluntary trial.
Affective	Loyalty to a liking: "I buy it because I like it."	Cognitively induced dissatisfaction. Enhanced liking for competitive brands, perhaps conveyed through imagery and association. Variety seeking and voluntary trial. Deteriorating performance.
Conative	Loyalty to an intention. "I'm committed to buying it."	Persuasive counterargumentative competitive messages. Induced trial (e.g. coupons, sampling, point-of-purchase promotions). Deteriorating performance
Action	Loyalty to action inertia, coupled with the overcoming of obstacles.	Induced unavailability (e.g. stock lifts – purchasing the entire inventory of a competitor's product from a merchant). Increased obstacles generally. Deteriorating performance.

Cognitive loyalty. In this first phase of loyalty the brand information available to consumers indicates that one is preferable over its alternatives. The loyalty is not very strong and the customer can be convinced to buy an other brand through product information. If satisfaction is processed (e.g. trash pickup, utility provision) it begins to take on affective overtones.

Affective loyalty. In this second phase a liking or attitude is developed toward the brand due to an accumulation of satisfying usage occasions. Here Oliver uses the confirmation disconfirmation paradigm known from satisfaction research (Oliver, 1980; Oliver, 1993; Oliver, 1997). Commitment at this phase is referred to as affective loyalty. Cognition is subject to counter arguments whereas affect is not as easily influenced. This type of loyalty is still subject to switching, thus, it would be desirable if customers were loyal at a deeper level of commitment. Satisfaction only cannot keep customers from buying different products (Jones and Sasser, 1995).

Conative loyalty. Conative loyalty is the next stage of loyalty development, which is influenced by related incidents of positive affect toward the brand and as a result leading to behavioral intention to use. Conation implies brand specific commitment to repurchase. In this stage at first appears deeply held commitment to buy as specified in the loyalty definition. The customer intends to rebuy, he or she has the motivation, but similar to any other motivation this desire may be an anticipated but unrealized action.

Action loyalty. For Oliver this stage is the connection between intention and action. The action control paradigm includes that a customer even overcomes obstacles to buy the desired product. This requires the highest level of commitment.

Oliver (1999, 36) identifies two main obstacles to loyalty - consumer idiosyncrasies and switching incentives. Marketing activities should address these two issues. Oliver's model can be applied for goods and services, with interpersonal factors appearing in a services setting that need to be acknowledged.

5.4.1.5 Measuring Loyalty

The high number of loyalty concepts lead to various operationalizations and measurement models. Oliver's (1999) approach for instance would require the inclusion of sustainers and vulnerabilities to enable a differentiation of the loyalty phases. He suggests a five point Likert scale with one item corresponding with each loyalty phase. The following table gives details on this operationalization.

Table 11: Measuring Loyalty According to Oliver (1997, 398)

Loyalty phase	Items
Cognitive	Brand X has more benefits than others in its class.
Affective	I have grown to like brand X more so than other brands.
Conative	I intend to continue buying brand X in the future.
Action	When I have a need for a product of this type, I buy only brand X.

Oliver (1997) also gives an overview of alternative operationalizations. In line with research and as already described the loyalty concept consists of attitudinal and behavioral aspects. Thus, both of them have to be included in the measurement instrument. There are two ways of measuring behavioral intentions, either ex post investigating repeat purchase, or ex ante investigating repurchase intentions. One limitation is that intentions do not always lead to the specified behavior. The attitudinal component can be measured directly (reports on attitudes) or indirectly (intention to recommend, price sensitivity).

Finally, concluding the discussion on loyalty the measurement in this research will use a two dimensional loyalty concept including behavioral and attitudinal components. The behavioral component will be measured by investigating the intended behavior and the past behavior. The attitudinal component is measured indirectly assessing the users' responsiveness to competitors' promotional efforts.

5.4.1.6 The Relationship between Customer Satisfaction and Loyalty

Literature pertaining to the relationship between customer satisfaction and loyalty can be organized in three categories:
1. Provision of empirical evidence of a positive relationship between customer satisfaction and loyalty without further elaboration.
2. Investigation of the functional form of the relationship between customer satisfaction and loyalty.
3. The examination of the effects of moderator variables on the relationship between the two constructs.

Research in the first stream has been based on the assumption of linear relationships (Homburg and Giering, 2001). However, researchers in the second category have provided theoretical and empirical support for a more complex, non-linear structure. Some suggest a convex relationship (Jones and Sasser, 1995), others a saddle curve shape of the relationship (Woodruff, Cadotte et al., 1983). Findings by Oliva et al. (1992) indicate that, depending on the magnitude of transaction costs, the relation between the two constructs can be both linear and nonlinear. The third and very limited group examines the existence of moderating factors on the relationship between satisfaction and loyalty (Homburg and Giering, 2001).

5.4.2 Value

Explanations of consumers' services purchases have circled around the relationship between service quality and purchase intentions (Cronin and Taylor, 1992; Parasuraman, Zeithaml et al., 1985; Parasuraman, Zeithaml et al., 1988; Parasuraman, Zeithaml et al., 1994b; Zeithaml, Berry et al., 1996). However, customers do not always buy the highest quality product nor do they always buy the lowest cost products. In this dilemma the value construct has been introduced to enhance the understanding of the importance of price and service quality. There have been various attempts to define value. Zeithaml (1988) identified four patterns of responses in an exploratory study, resulting in four potential definitions of value:
- Value is low price.
- Value is whatever one wants in a product.
- Value is the quality that the consumer receives for the price paid.
- Value is what consumers get for what they give.

While each of these definitions has merit, the vast majority of past research has focused on the fourth meaning of value (Bojanic 1996;

Caruana, Money, and Berthon 2000; Zeithaml 1985). As Zeithaml suggested as a tradeoff between relevant *"gives"* and *"gets"*.

Specifically Zeithaml defines perceived value as *"the customer's overall assessment of the utility of a service based on perceptions of value"* (Zeithaml and Bitner, 1996, 501).

Review of perceived value literature leads to two main perspectives that have been used to model customers' value perceptions: utilitarian and behavioral (Jayanti and Fosh, 1996).

5.4.2.1 Utilitarian Approach to Perceived Value

Some authors argue that consumers' value perceptions result from comparing different prices of products. This includes the selling price, advertised reference price and internal reference price (Monroe, 1990; Thaler, 1985). According to Monroe (1990) advertised selling prices and advertised reference prices influence potential customers' internal reference prices. This suggests that customers reflect relevant information and then form price expectations and perceptions on value. Perceived value consequently is a combination of acquisition value and transaction value of that product.

Parasuraman and Grewal (2000) identified four types of value in the literature (Grewal, Monroe et al., 1998; Woodruff, 1997):

- *Acquisition value*: This is the benefit related to monetary cost the buyers believe they are getting by acquiring a product or service.
- *Transaction value*: This relates to the pleasure of getting a good deal.
- *In-use value*: This is the utility deriving from using the service/product.
- *Redemption value*: This is the residual benefit at the time of trade-in or end of life for products or the time of termination for services.

All these conceptualizations imply a dynamic perspective of perceived value and the importance of those components may change over time. Earlier work shows that antecedents of value changes over time and at the different stages of a customers' lifecycle (Parasuraman, 1997; Woodruff, 1997).

Grewal et al. (1998) develop a model that shows the effects of advertised selling and reference prices on buyers' internal reference prics, perceptions of quality, acquisition value, transactionvalue, purchase, and search intentions. Grewal et al.'s (1998) conceptualization of perceived transaction value combines previous conceptualizations and suggests that it is a *"psychological satisfaction or pleasure obtained from taking advantage of the financial terms of the price deal"*. Grewal et al. (1998) argue that buyers' perceptions of price deals are based on comparisons of different price structures which are included in the model. They are: advertised selling price, advertised reference price, internal reference price. If the comparison yields a positive result customers are satisfied which in turn is the transaction value of a product.

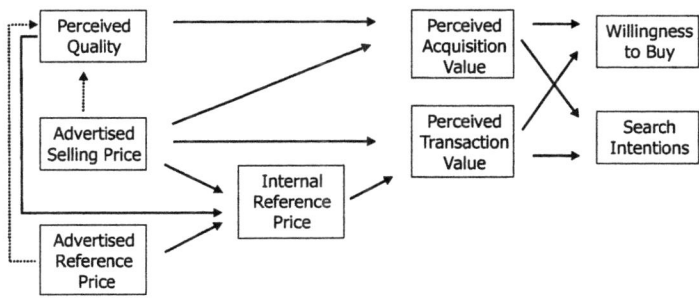

Figure 17: Effects of Price Comparison on Perceptions of Value (Grewal et al. 1998)

Conceptualizing perceived value exclusively based on price deals is an important but insufficient approach. Customers also take other product attributes into consideration. Thus, other models have been suggested to explain perceived value.

5.4.2.2 Behavioral Approach to Perceived Value

Zeithaml's (1988) model of perceived value incorporates psychological antecedents and higher level attributes. It is a good starting point since many researchers considered it for their own conceptualizations.

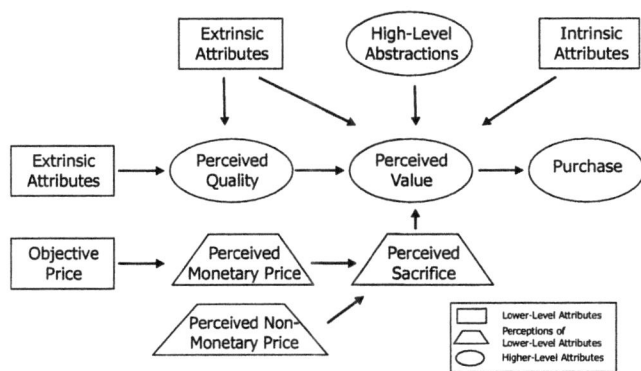

Figure 18: Means-End Model Relating Price, Quality, and Value (Zeithaml, 1988)

The above model indicates that quality perceptions are reached through evaluations of product attributes, which also lead to overall value judgments. Therefore, Zeithaml (1988) suggests that the formation of quality and value perceptions occur in a means-end way. Literature review shows that the means-end approach was used to demonstrate how means as objects or activities are connected to ends (i.e. desired end states or values) (Gutman, 1982). There are three levels of abstraction in this means-end chain, from objects to values, that consumer's memorize in the end. First, attributes, representing the lowest level. Then, the consequences of those attributes witch can be quality and value judgments in this case.

Woodruff and Gardial (1996) and Woodruff (1997) based their wok on Zeithaml (1988) and suggest a similar explanation based on the means-end considerations. However, they propose their own model of consumer value to explain the three levels of abstraction also used by Zeithaml (1988). This is shown in the following figure.

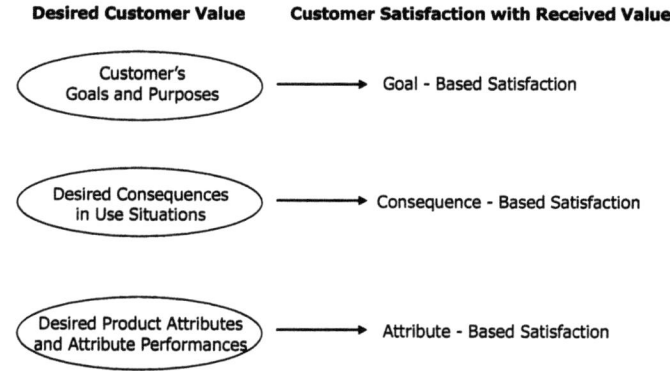

Figure 19: Consumer Value Hierarchy Model (Woodruff, 1997)

Woodruff (1997) argues that customer value should be conceptualized at three levels, namely the levels of attributes, consequences, and goals. These three levels are incorporated into the *'Customer Value Hierarchy Model'* depicted above. In this hierarchy consumers learn to relate the specific attributes to consequences and end-goals.

Attribute based satisfaction are the first level. Initially, at the pre-purchase stage, customers are concerned to identify and assess the products as bundles of attributes.

Consequences are the next step. After buying/ using a product the customer learns how well it performs. The consequences may be positive of negative.

Goals are at the highest level. Consumers at this stage understand the contribution of the product to the achievement of their goals.

Woodruff and Gardial (1996) argue that customers want to achieve their end goals in every single consumption. The two columns of figure 19 (desired value and received value) distinguish between customer expectations (desired value) and customer-perceived performance (received value).

Other perspectives on value perceptions can be referred to as typologies. Holbrook (1999) for instance defines consumer value as *"an interactive relativistic preference experience"*. His value typology consists of various dimensions. Based on the definition the dimensions of value are: efficiency, play, excellence, aesthetics, status, ethics, esteem, and spirituality. Holbrook includes extrinsic versus intrinsic, active versus reactive, and self-oriented versus other-oriented factors to build an extensive model.

		Extrinsic	**Intrinsic**
Self-oriented	Active	**Efficiency** (Convenience)	**Play** (Fun)
	Reactive	**Excellence** (Quality)	**Aesthetics** (Beauty)
Other-oriented	Active	**Status** (Success, Impression, Management)	**Ethics** (Virtue, Justice, Morality)
	Reactive	**Esteem** (Reputation, Materialism, Possessions)	**Spirituality** (Faith, Ecstasy, Sacredness, Magic)

Figure 20: Holbrook's Typology of Consumer Value (Holbrook, 1999, 12)

Similar to Holbrook (1996) Sheth et al.'s (1991) conceptualization is also benefit-driven. They provide five different values that drive consumer choice behavior as motivational forces. Those consumer value types explain the reasons behind consumers' product choice. According to Sheth et al. (1991) consumers buy products to reach one of the following value types:
- functional value (attribute related, utilitarian benefit)
- social value (social or symbolic benefits)
- emotional value (experimental or emotional benefits)
- epistemic value (curiosity driven benefits)
- conditional value (situation specific benefits e.g., Christmas Valentines day etc.)

Both, Holbrook's (1996) and Sheth et al.'s (1991) models only explain value by consumption benefits. The shortcoming is that no cost involved is considered.

5.5 Discussion of the Relevant Models and Constructs

Researchers deal with service adoption and service quality in various publications. This chapter summarizes and discusses important factors and findings from previous quantitative surveys.

First of all perceptions rather than objective technology attributes have been found to be more relevant to technology acceptance decision making (Moore and Benbasat, 1991). In Moore and Benbasat's (1991) model Rogers' complexity construct was renamed and called ease of use, consistent with Davis (1989), reflecting the dominant measurement paradigm in ICT research. They also developed the image construct which is comparable to social norm and one could argue that image was included by Rogers' in his definition of the construct relative advantage. Rogers (1995, 204) discusses the importance of utilizing consistent instruments or measurements of innovation attributes to contribute to innovation diffusion research.

A myriad of such instruments, often with very little difference, just renaming concepts and strongly drawing on Rogers' considerations emerged. Rogers draws attention to the fact that effort has been spent studying people related differences in innovativeness while relatively little effort has been devoted to analyzing innovation differences. In other words, differences on how attributes of innovations affect the rate of adoption. Concluding this line of arguments, in this research the attributes of an innovation will be evaluated and additionally the level of innovativeness of individual's will be taken into account as a moderating variable.

For this survey the most important conclusion from the above described quality models is the differentiation between perception and expectation. One could argue that the quality perception of a mobile service user is determined by his expectations. Yet, researchers found that with new technology products customers do not know how well a new service could work and, thus, do not have precise expectations on its performance. Therefore, perceptions on the quality are the main object of interest.

Factors of Lociano et al. (2002) webqual model show constructs that also appeared in most adoption models and also can be drawn from the diffusion of innovations theory.

One of the main differences and achievements of Internet quality models is the acknowledgement of enjoyment or fun factors (Van der Heijden, 2003; Van der Heijden, 2004).

Furthermore, the perceived value of a service and its impact on loyalty is considered. In this study loyalty is a construct with attitudinal and behavioral components.

The following table gives an overview of relevant adoption and service quality surveys, including the method employed, the model used and additional constructs as well as the supported structural relations and major findings.

The review shows that most surveys empirically supported the models described in the previous chapters. Some extended the existing models to include: perceived enjoyment, perceived attractiveness, image, visibility, result demonstrability, gender, social presence/information richness, performance, fun, self-efficacy, novelty seeking, need for interaction, self-consciousness, perceived waiting time, social anxiety. The following moderators were identified: gender, age, experience, voluntariness of use.

Reviewing the models applied in mobile services research yielded the following extensions of traditional models: user friendliness, self control, self-expressiveness, connection speed, service costs, personal innovativeness, perceived cost, social influence, perceived risk, relative advantage, compatibility, communicability, critical mass, navigation, payment options, perceived risk, cost, compatibility, fun, consumer visual orientation, Internet device. The review lead to the following moderator variables: age, computer skills, mobile technology readiness, social influence.

Further details can be found in the table below.

Table 12: Diffusion, Adoption and Service Quality Models in IS and Mobile Services Research

Model/ Theory	Additional Constructs	Supported Causal Relations	Findings
Barnes and Vidgen (2002): Qualitative and quantitative Surveys			
WebQual		Usability, Design, Information, Trust, Empathy $\rightarrow$ Q	WebQual is a method for assessing the quality of Web sites, developed through application in various domains. WebQual is grounded in the impression of web users.
Cronin, Brady and Hult (2000): Survey, n=1844			
Quality, Value, Satisfaction		$Q \rightarrow I, V, Sat$ $V \rightarrow I, Sat$ $Sat \rightarrow I$ $Sacrifice \rightarrow V$	The study investigates the effects of quality, satisfaction and value on consumer's behavioral intentions. The article compares competing models. It has to be noted that service quality, satisfaction and value are all directly related to behavioral intention.

Davis et al. (1989): Experiment			
TRA, TAM		EOU →U, A U → A, I A → I I → Use	EOU is secondary and acts through U. Attitudes have little impact mediating between perceptions and intention to use.
Davis (1989): Survey, n=152			
TAM		U → Use, EOU →Use	TAM fully mediated the effects of system characteristics on use behavior, accounting for 36% of the variance in use. Usefulness is 50% more influential than ease.
Dabholkar and Bagozzi (2002): Survey, experimental design, n=392			
TAM	Performance, Fun, self-efficacy, novelty seeking, need for interaction, self-consciousness, perceived waiting time, social anxiety	A → I PEN → A EOU → A Performance → A Moderators: self-efficacy, novelty seeking, need for interaction, self-consciousness, perceived waiting time, social anxiety	This survey investigates the moderating effects of consumer traits and situational factors on the relationships of a core attitudinal model based on TAM. The results lend support to the hypothesized moderating effects.
DeLone and McLean (1992): Exploratory study			
I/S Success		System qual., information qual → use, user satisfaction → Individual impact → organizational impact	The authors introduce a taxonomy for IS success, integrating the diverse past research in this field. The main aspects are drawn into a descriptive model and the implications for future IS research are discussed.
Gefen and Straub (1997): Survey, n=392			
TAM	Gender, Social presence/ information richness (SPIR)	Gender → U, EOU, SPIR, Usage SPIR → U U → Usage EOU → Usage	The study findings indicate that women and men differ in their perceptions but not in use. This suggests that gender should be included in diffusion models along with other cultural effects
Karahanna et al. (1999): Survey, n=268			
IDT, TRA, TPB	Image, visibility, result demonstrability	A → I SN → I EOU, Image, U, Visibility, Result Demonstrability, Trialability → A Top Management, Supervisor, Peers, MIS Department, Local Computer Specialist, Friends → SN	Pre-adoption attitude is based on perceptions of usefulness, ease of use, result demonstrability, visibility and trialability. Post-adoption attitude is only based on instrumental beliefs of usefulness and perceptions of image enhancements.

Loiacono, Watson and Goodhue (2000): Survey n=510, 336, 311			
WebQual		U, EOU, PEN, complimentary relationship → Q	Development of a measurement instrument for Web site quality. 12 core dimensions with 4 umbrella items: usefulness, ease of use, entertainment and complimentary relationship.
Mathieson (1991): Experiment			
TPB, TAM		Same as Davis (1989)	Both TAM and TPB predict well. TAM is easier to apply; TPB provides more specific information for developers.
Taylor and Todd (1995a): Survey, n=786			
TAM, TPB		U → A EOU → A, U A → I SN → I PBC → I, Use I → Use	The model consisting of TAM and TPB constructs explains for both, experienced and inexperienced users. The link between behavioral intention and behavior is stronger for experienced users. Antecedent variables predict inexperienced user's intentions better.
Parasuraman et al. (2005): Survey, Measurement instrument development			
E-S-QUAL		Q → V Q, V → Loyalty intention	The E-S-QUAL scale consists of 22 items falling into 4 factors: Efficiency, fulfillment, system availability, and privacy. The E-RecS-QUAL scale (for customers with non-routine encounters with the sites) contains 11 items and three dimensions: responsiveness, compensation, and contact.
Teo et al. (1999): Web based survey, n=1370			
TAM	Perceived enjoyment	EOU →U, Usage, PEN U → Use PEN → Use	
Van der Heijden (2004): Survey, n=1144			
TAM	Perceived enjoyment	U → I PEN → I EOU → PEN, U	The survey supports the hypotheses that perceived enjoyment and perceived ease of use are stronger determinants of intentions to use than perceived usefulness.
Van der Heijden (2003): Web based survey, n=828			
TAM	Perceived Attractiveness, Perceived enjoyment	EOU → U, A U → I, A P. attractiveness → U, EOU, PEN PEN → A, I A → I I → Use	Intention is most influenced by attitude, less by enjoyment and usefulness. Visual attractiveness is an important construct.

Venkatesh et al. (2003): Survey, n=215				
UTAUT	Moderator: Gender, age, experience, voluntariness of use	EOU/Effort Expectancy → I U/Performance Expectancy → I PBC/Facilitating conditions → I SN → I I → Use	This is a new model based on previous models/theories with the attempt to build a unified model. This measured well with respect to the sample chosen but further model tests are needed.	
Wolfinbarger and Gilly (2003): Focus groups, Tasks, Survey				
eTailQ	Quality, Fulfillment/Reliability, Website Design, Privacy/Security, Customer Service	Fulfillment, Website Design, Privacy/Security, Customer Service → Quality	The analysis suggests that Quality, Fulfillment/Reliability, Website Design, Privacy/Security and Customer Service are strong predictors for customer judgments of quality, satisfaction, customer loyalty and attitudes toward the website.	

Mobile Services Area			
Bruner and Kumar (2005): Survey, n=212			
TAM	Fun, Consumer visual orientation, Internet device	U → A EOU → U, PEN A → I PEN → A Cons. Visual orientation → EOU Internet device → EOU. PEN	TAM was extended by utilitarian and hedonic aspects. Fun contributes more to attitude than expected and visually oriented people will adopt more than less visually oriented ones.
Hung et al. (2003): Survey, n=267			
TPB, IDT	Connection speed, service costs, user satisfaction, personal innovativeness	Connection speed → A User Satisfaction → A INN → A EOU → A U → A A → I Peers → SN SN → I Self Efficacy → PBC PBC → Use	The total variance explained in actual WAP usage is rather low, most respondents have no experience. All in all, the model fits well.
Kleijnen, Wetzels et al. (2004): Survey, n=99			
IDT	Perceived risk, relative advantage, compatibility, communicability, critical mass, navigation, payment options	Cluster profiling variables (innovativeness, leadership, Internet usage) lead to three clusters	Perceived risk plays a critical role in the adoption process, followed by complexity and compatibility. Also three segments are identified *"Value Seekers", "Game Players"* and *"Risk Avoiders"*.

Kleijnen et al. (2004): Survey, n=105			
TAM	Perceived system quality, perceived cost, social influence. Moderators: Age, computer skills, mobile technology readiness, social influence	U → A EOU → U A → I Perceived system quality → A SN → I	From the three constructs added to the TAM model, system quality and social influence displayed significant effects. All of the moderating variables proved to be relevant in the context presented.
Pedersen (2003): Survey, n=190			
TRA, TAM, TPB	User friendliness, self control	U → A SN → I, A User friendly, External influence → U A → I PBC → I, I → Use Self control, interpersonal infl. → SN Self efficacy, facilitating conditions → PBC	The survey shows that a model integrating various concepts such as TPB, PBC and TAM measure well. Adding behavioral control to the model increased the explanatory power where as adding subjective norm did not. The simple TAM only explained 30% of variance in intention to use and 17% of variance in actual use. The complex model explained 49% of the variance in intention to use. Subjective norm combined with behavioral control improves model fit and adds to explanatory power.
Pedersen and Herbjorn (2003): Web based survey, n=452			
TAM	Self-expressiveness	A → I U → A, I EOU → U, A Self-expressiveness → A, I, U	When self-expressiveness is removed from the model the model fit is reduced. Additionally the explained variance in intention to use is also reduced from 58% to 50%.
Wu and Wang (2005): Survey, n=310			
TAM2, IDT	Perceived risk, cost, compatibility	EOU → U, I I → Use U → I Compatibility → I, U Cost → I Perceived risk → I	All variables except perceived ease of use affected user's behavioral intention. A puzzling finding is the positive influence of perceived risk on behavioral intention. The most important determinant for behavioral intention to use is compatibility.

Legend: A=attitude, U=usefulness, I=behavioral intention, EOU=ease of use, PEN=perceived enjoyment/fun, INN=innovativeness, SN=subjective norm, PBC=perceived behavioral control, Q=quality, V=value, Sat=satisfaction, SPIR=social presence & information richness. For an additional review of related TAM, TPB, TRA, IDT, UTAUT, MM, C-TAM/TPB, MPCU etc. studies please refer to Lederer et al. (2000) Legris et al. (2003) and Gefen and Straub (2000).

5.6 Causal Modeling

"Structural equation modeling (SEM) is a statistical methodology that takes a confirmatory (i.e. hypothesis-testing) approach to the analysis of a structural theory bearing some phenomenon." (Byrne, 2001, 3)

However, Pearl (2000) takes a more critical approach indicating that the hypothesis test should not be the main emphasis. Pearl's (2000) book discusses causality and structural models in social science arguing that the most distinctive capabilities of SEM are currently ill understood and underutilized. He uses graphical models and the logic of intervention to alleviate the current difficulties. The ambitious goal is to reinstate the causal interpretation of SEM.

Causal analysis, often also called structural equation modeling, covariance structure analysis, structural equation methodology, causal modeling, LISREL (Linear Structural Relations) -Approach, or latent variable modeling is established in the recent empirical marketing science (Homburg and Hildebrandt, 1998).

Accordingly, SEM falls into the category of second generation multivariate methods typically conveying causal processes. These processes under study are represented by a series of structural (i.e. regression) equations. The main difference between the second and first generation (e.g. multidimensional scaling, factor analysis) of multivariate analysis are the following (Hulland, Chow et al., 1996):
1. The inclusion of measurement errors
2. The possibility to include abstract and latent constructs
3. The opportunity to combine theory and data and confront theory with data

The model of representing a set of linear relations devised by Joreskog incorporates both constructs or latent variables and their (multiple) indicators. It combines the causal approach with the powerful measurement technique of ML factor analysis. Therefore, according to Mazanec (1982) a latent variable-multiple indicator model can only be confimed or rejected in its entirety inclusive of all measurement assumptions.

Covariance matrices depict the associations of observed variables, leading to the explanation of relations of a smaller number of underlying constructs. The fact that causal analysis enables the modeling and estimation of complex structures of dependence simultaneously is useful in the behavioral sciences. In this discipline researchers often study theoretical constructs that cannot be observed directly (Byrne, 2001, 4). This requirement is often recognized in consumer behavior and adoption theory, where constructs tend to be complex and, by definition, not directly observable. Consequently, many authors used structural equation modeling (Aijzen, 1991; Ajzen, 2001; Ajzen and Fishbein, 1980; Davis, 1989; Davis, Bagozzi et al., 1989; Davis, Bagozzi et al., 1992; Fishbein and Ajzen, 1975; Gefen and Straub, 2000; Kleijnen, Ruyter et al., 2002; Pedersen, 2003; Pedersen and Herbjorn, 2003; Pedersen, Leif et al., 2002; Straub, Keil et al., 1997; Varshney, 2003; Varshney and Vetter, 2001; Venkatesh, 2000; Venkatesh and Davis, 2000; Venkatesh, Morris et al., 2003).

Also in service quality and consumer behavior SEM is frequently employed (Ajzen, 2001; Bagozzi, 1980; Bagozzi, 1994; Baumgartner and Homburg, 1996; Bolton and Drew, 1991; Cronin, Brady et al., 1997; Cronin, Brady et al., 2000; Grewal, Monroe et al., 1998; Homburg and Baumgartner, 1998; Homburg and Giering, 2001; Matzler, Bailom et al., 2004; Oliver, 1993; Oliver, 1999; Parasuraman, Zeithaml et al., 1994a; Steenkamp and Baumgartner, 2000; Steenkamp and Baumgartner, 2001; Zeithaml, 1988).

5.6.1 Causal Models

A structural model consists of various variables which can be graphically represented. The following figure depicts these graphic symbols. The representation of a structural model via mathematical formulas follows later in this section.

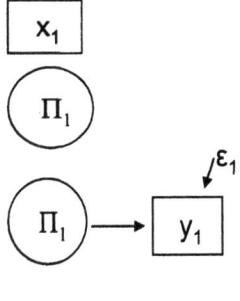

A Rectangular or square box signifies an observed or manifest variable.

A circle or ellipse signifies an unobserved or latent variable.

An unenclosed variable signifies a disturbance term (error in either equation of measurement). A straight arrow signifies the assumption that a variable at the base of the arrow "causes" the variable at the head of the arrow.

A curved two-headed arrow signifies an unanalyzed association between two variables.

Two straight single-headed arrows connecting two variables signify feedback relations or reciprocal causation.

Figure 21: Primary Symbols Used in Path Analysis (Bollen, 1989, 33)

A major characteristic of causal models is the differentiation of observable (manifest) and latent variables. The latter are more complex

89

constructs which cannot be observed directly. Latent variables are commonly called *factors* and observed (or manifest) variables are named *indicators*. Above that one should distinguish between exogenous latent variables and endogenous latent variables.

"*Exogenous latent variables are synonymous with independent variables; they 'cause' fluctuations in the values of other latent variables in the model. ... Endogenous latent variables are synonymous with dependent variables and, as such, are influenced by the exogenous variables in the model, either directly or indirectly.*" (Byrne, 2001)

The structural model shows the hypothesized relations between the latent variables and the manifest ones with the help of factor analysis. One can distinguish between the measurement model for the latent exogenous and the latent endogenous variables.

The following figure shows a classical causal model consisting of a structural equation model, an exogenous, and an endogenous factor model.

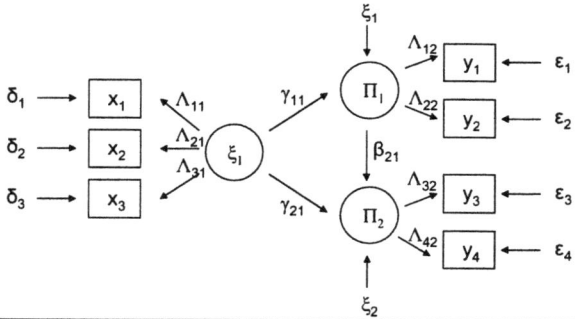

Measurement Model	Structural Model	Measurement Model
$x_1 = \Lambda_{11} \xi_1 + \delta_1$	$\Pi_1 = \gamma_{11} \xi_1 + \zeta_1$	$y_1 = \Lambda_{12} \Pi_1 + \varepsilon_1$
$x_2 = \Lambda_{21} \xi_1 + \delta_2$	$\Pi_2 = \gamma_{21} \xi_1 + \beta_{21} \Pi_1 \zeta_2$	$y_2 = \Lambda_{22} \Pi_1 + \varepsilon_2$
$x_3 = \Lambda_{31} \xi_1 + \delta_3$		$y_3 = \Lambda_{32} \Pi_1 + \varepsilon_3$
		$y_4 = \Lambda_{42} \Pi_1 + \varepsilon_4$

Figure 22: Causal Model (Homburg and Hildebrandt, 1998, 19)

The covariance or correlation matrix is the beginning of the analysis. Covariances and correlations are calculated based on the indicator variables from the measurement model. Next to that an estimation of the relationships between latent and indicator variables and between the exogenous and endogenous variables is possible (Backhaus, Erichson et al., 2000). The following table summarizes the terminology.

Table 13: Notation for Latent and Measurement Model (Bollen, 1989, 14)

Symbol	Name	Dimension	Definition
Variable			
η	Eta	m*1	Latent endogenous variables
ξ	Ksi	n*1	Latent exogenous variables
Z	Zeta	m*1	Latent errors in equations
X		q*1	Observed indicators for Ksi
Y		p*1	Observed indicators for Eta
Δ	Delta	q*1	Measurement errors for x
E	Epsilon	p*1	Measurement errors for y
Coefficients			
B	Beta	m*m	Coefficient matrix for latent endogenous variables
Γ	Gamma	m*n	Coefficient matrix for latent exogenous variables
Λ_x	Lambda x	q*n	Coefficients relating x to Ksi
Λ_y	Lambda y	p*m	Coefficients relating y to Eta
Covariance Matrices			
Φ	Phi	n*n	Covariance matrix of Ksi
Ψ	Psi	m*m	Covariance matrix of Zeta
Θ_δ	Theta-Delta	q*q	Covariance matrix of Delta
Θ_ϵ	Theta-Epsilon	p*p	Covariance matrix of Epsilon

5.6.2 Usage Areas for Causal Models

Generally, causal analysis is considered as a tool for the test of hypotheses; in some cases it is a tool for explorative research. Homburg and Hildebrand (1998) distinguish between four use cases:
1. Construct validation
2. Test of hypotheses
3. Group comparisons of model structures
4. Exploration of structures

These usage cases have been widely accepted, however, one needs to consider Pearl's (2000) work on causality providing a comprehensive exposition of modern analysis of causation. He argues that structural equations are often interpreted as carriers of probabilistic information instead of carriers of substantial causal information. He condends that SEM focuses too much on model fitting while issues regarding the meaning and usage of SEM's models are subject of confusion and controversy.

The validation of constructs is by far the most frequent use of causal models and became more widely accepted in science than correlative methods. Validity is mostly tested with convergent and discriminant validity. Baumgartner and Homburg (1996) found that nearly half of the applications of causal modeling deal with construct validation (Homburg and Hildebrandt, 1998).

Secondly, causal modeling is frequently used for hypotheses testing. For complex causal model structures, which are often observed in business administration and are suggested in real life, causal modeling is a strong tool. Thus, the above statement that half of all usage cases of causal models would be limited to an analysis of the measurement model not taking the relations of constructs in the structural model into account.

By separately testing the measurement and the structural model much can be gained in theory testing and the assessment of construct validity. This is an exploratory exercise. The measurement model with the structural model enables confirmatory assessment of construct validity (Bentler and Chou, 1987) and the measurement model only provides a confirmatory assessment of convergent validity and discriminant validity (Campbell and Fiske, 1959). If the discriminant and convergent validity are both satisfying the test of the structural model then constitutes a confirmatory assessment of the nomological validity (Campbell and Fiske, 1959).

Concerning this research, causal analysis is a confirmatory, hypotheses testing instrument.

5.6.3 Steps for Structural Equation Modeling

To analyze established models using causal analysis a number of steps proved successful. This approach is suggested by various authors (Backhaus, Erichson et al., 2000; Bollen, 1989; Bollen and Long, 1993; Homburg and Hildebrandt, 1998; Kelloway, 1998; Schumacker and Lomax, 1996).

The main stages characteristic of most applications of structural equation modeling, which are described into more detail in the following sections are model specification, model identification, parameter estimation, and model evaluation.

5.6.3.1 Model Specification

The specification of a general structural equation model involves three distinct tasks (Mueller, 1996):
1. A specific structure between latent exogenous and endogenous constructs must be hypothesized
2. The measurement of the exogenous latent variables has to be decided
3. Determination of a measurement model for endogenous latent constructs

The propositions for the composition of the a priori model are most frequently drawn from previous research (Bollen and Long, 1993). The purpose of the model is to explain why variables correlate in a certain way.

With regard to the measurement of the factors, in practice there are sometimes only single indicators available that do not perfectly estimate the constructs. Ideally the researcher would have an independent measure from previous research which is often not available. Also, the choice of values can be arbitrary.

Non convergence or improper solutions can also occur with small sample sizes. Anderson and Gerbing (1984) found that a sample size of 150 are sufficient to obtain a converged and proper solution for models with three or more indicators per factor. If there are only two indicators per factor bigger samples may be needed.

5.6.3.2 Model Identification

While the estimation of models may sound simple in theory several problems may be encountered before estimates are obtained. The identification status (i.e. under-, just-, or overidentified) is difficult to prove mathematically and the identification problem must be dealt with.

The process of obtaining empirical estimates of model parameters of a model is either solving a set of equations or minimizing a function. A model with insufficient information to obtain an estimate for each and every parameter is *nonidentified*. A model with an equal number of equations and unknown parameters is *just identified* (Kelloway, 1998). If there are more equations than parameters to be estimated the model is *overidentified*.

The simple prerequisite to have as many equations as unknowns may be too general; certain parts of the model may be non- or overidentified. Therefore, several indicators for each latent variable should be chosen. The statistical power may also suffer if this criterion is not met (Kelloway, 1998).

5.6.3.3 Parameter Estimates

The model and the unknown parameters generate an implied covariance matrix. This should be similar to the observed covariance matrix. Now the parameters that make the difference as small as possible are to be found. A model may be estimated by several different methods or in other words by using different estimators.

The by far most used fitting function is based on the maximum likelihood (ML) considerations followed by generalized least squares (GLS) (Anderson and Gerbing, 1988).

The maximum likelihood estimator may be that popular as it is included in the most common software packages as LISREL and Amos, even though those packages also include other estimators. ML is efficient with large samples (Bollen, 1989) and when the researcher is willing to assume (or show) that the observed variables are multivariate normal. Only then the chi-square test is reasonable. Since these underying assumptions cannot be verified for this study a different estimator provided in Mplus is chosen.

Opposed to ML and GLS, which are full information methods, ordinary least squares is known as partial information technique. With partial information techniques each path is estimated independently of the others.

To solve the multivariate normality problem Browne (1984) introduced a distribution free estimation technique (Asymptotic Distribution Free estimator, ADF). However, according to Chou and Bentler (1995), big sample sizes are needed and the technique is computationally cumbersome.

The most used ones, ML and GLS are both recommended for theory testing and development (Jöreskog and Wold, 1982), unweighted least squares, generalized least squares or partial least squares are recommended for application and prediction (Jöreskog and Wold, 1982). For a discussion on the above mentioned estimators the reader may refer to Anderson and Gerbing (1988) for further details.

Since in this project the software package MPlus is used for data analysis, the Muthén estimator is employed for analysis. The used software package is a second generation tool in SEM and allows for estimation of thresholds when categorical data is used, which is the case in this survey. The Estimator used in this survey is further explained in chapter 5.6.4.2 on the estimators offered in MPlus.

Finally, the underlying assumptions on which the different estimators are based should be carefully considered before choosing one or the other.

5.6.3.4 Model Evaluation Measures

Structural equation modeling determines if a theoretical model successfully shows the actual relationships observed in the sample data. The output of the analysis provides indices that demonstrate whether the model conforms to the data and also demonstrates significance tests for specific causal paths. The researchers postulate a statistical model based on their knowledge of the related theory or/and on empirical research in the area of study. Once the model is specified, the researcher tests its plausibility based on sample data comprising all observed variables in the model. The primary task in this model-testing procedure is to determine the goodness of fit between the hypothesized model and the sample data (Byrne, 2001).

The discussion about Goodness of Fit Indices and how to assess validity of a structural equation model has been tremendous. Most of the software packages utilized for structural equation modeling calculate over 20 different indices. Among the most commonly used are the
- Chi-square statistic
- Goodness-of-fit and adjusted goodness-of-fit (Jöreskog and Sörbom, 1981)
- Normed and nonnormed fit indices (Bentler and Bonett, 1980)
- Normed comparative fit index and nonnomed fit index (Bentler, 1990)
- Parsimonious goodness-of-fit and parsimonious normed fit indices (Mulaik, James et al., 1989)

Each index has advantages and disadvantages depending on the underlying assumptions (e.g. they all depend on the multivariate normality assumption) the researcher has to be aware of before rejecting or accepting a model. Mueller (1996) points out some straightforward and valuable aid in identifying data-model inconsistencies – scrutinizing the individual parameter estimates. An understanding of the substantive theory and hypothesized model combined with statistical knowledge go a long way in assessing the adequacy of the proposed structure. Details on fit indicators can be found in: Anderson and Gerbing (1984), Arbuckle and Wothke (1999), Bentler (1990), Bentler and Bonett (1980), Browne and Cudeck (1993), Hu and Bentler (1995), Jöreskog and Sörbom (1981), Mulaik, James et al. (1989), Muthén and Muthén (1998). Details for measures of component fit can be found in: Gerbing and Anderson (1988), Fornell and Larcker (1981), Bagozzi and Yi (1988) and Homburg and Baumgartner (1998).

Table 14: Fit Indices (Hair, 1995; Hu and Bentler 1995)

Goodness-of-fit Measures	Levels of Acceptable Fit
Overall Fit	
Root mean square error of approximation (RMSEA)	$\leq 0.05 / \leq 0.08$
Goodness-of-fit index (GFI)	≥ 0.09
$\dfrac{\chi^2}{df}$	≤ 2.5
Adjusted goodness-of-fit index (AGFI)	≥ 0.90
Comparative Fit Index (CFI)	≥ 0.90
Standardized Root Mean Square Residuals (SRMR)	$\leq .08$
Weighted Root Mean Square Residual (WRMR)	$\leq .09$ or close to 1
Structural Model	
Squared Multiple Correlation for each endogenous latent variable	(≥ 0.40)
Measurement Model	
Construct Reliability	≥ 0.60
Average Variance Extracted	≥ 0.50

5.6.4 M-Plus

Since the introduction of the LISREL approach, various software packages have been developed for the calculation of structural equation models. Among the best known are LISREL/PRELIS, EQS, AMOS, Mx and LISCOMP. Most of these do not accomodate categorical data. Mplus, LISCOMP's successor, was developed by Bengt Muthén in (1998) to estimate models including categorical and binary data. Mplus differs from LISCOMP with regard to three issues (Maydeu-Olivares, 2000). These three extensions now allow the estimation of:
1. Models with categorical latent variables
2. Models for continuous dependent variables that contain data missing completely at random (MCAR) and missing at random (MAR)
3. Models for two level (disaggregated) data obtained under complex sampling.

Maydeu-Olivares (2000) provides a detailed paper including the advantages and disadvantages of Mplus over LISCOMP.

Mplus allows various model classes; among those are multivariate regression analysis, path analysis, explorative and confirmatory factor analysis, SEM, latent class analysis and Monte Carlo simulation (Muthén and Muthén, 1998).

5.6.4.1 Particularities Estimating Models with Categorical Data

In Mplus observed variables can be measured on a continuous or categorical scale. Observed categorical variables include dichotomous (binary) and ordered categorical (polytomous) variables. Either unordered categorical or continuous is the measurement scale of latent variables. When categorical data is used there are some specific characteristics to consider: Thresholds instead of means are estimated for dependent categorical observed variables. Mplus only allows a maximum of ten categories; this has to be considered when specifying the survey design. Correlation matrices are analyzed instead of covariance matrices.

5.6.4.2 Estimators in Mplus

Mplus has five estimator choices:
- Maximum likelihood (ML),
- Maximum likelihood with robust standard errors and chi-square (MLM, MLMV),
- Generalized least squares (GLS), and
- Weighted least squares (WLS) also referred to as ADF.

When at least one factor indicator or observed variable is categorical, Mplus has four estimator choices:
- Weighted least squares (WLS),
- Robust weighted least squares (WLSM, WLSMV), and
- Unweighted least squares (ULS).

The WLS-estimators were developed by Muthén and Muthén (1998). WLSMV uses the diagonal of the weight matrix in the estimation whereas WLS uses the full weight matrix. WLS and WLSMV use the full weight matrix to compute standard errors and chi-square. Neither estimator uses a fitting function to minimize residuals. The WLSMV estimator however, uses the diagonal weight matrix to get the estimates, thus, the residuals tend to be closer to zero than using the WLS estimator.

This estimator is always used for modeling with categorical data and will therefore be used for analyses in this project. More details on the Muthén estimator are offered at www.statmodel.com where one also finds a list of related technical papers.

5.6.4.3 Fit Indices for Categorical Variables

Muthén and Muthén (1998) advise against the use of most fit indices for categorical data. *"The draw back is that little is known about how to use … (different fit indices) … for categorical outcomes in practice"* (Mplus, 2000). According to the authors only RMSEA, CFI and TLI are applicable for categorical data. However, even concerning those they advise: *"As far as we know, there have been no published studies of the behavior of RMSEA, CFI or TLI for categorical outcomes. Our very limited studies of RMSEA found that it does not work as well for categorical outcomes as for continuous."* (Mplus, 2000). Muthén and Muthén encourage studies in this area. Other authors do not have any reservations using fit indices such as the GFI that can be calculated theoretically.

6 Hypotheses Development and Model Specification

A cross disciplinary integration of theories may improve the understanding of basic mechanisms of individual's usage of mobile services. In this section integration is exemplified by suggesting how traditional models in information system research and consumer behavior may be modified and extended when applied to study the adoption of mobile services.

In discussing desirable characteristics of mobile services experts and users mentioned a high range of variables ranging from concrete cues (e.g. easy interaction with the service, easy to understand, fun to use) to more general perceptual attributes (e.g. perceived transaction speed, perceived security of interaction) to higher-order abstractions (perceived value, perceived quality). This is in line with the reasoning of Parasuraman, Zeithaml and Malhotra (2005) and the assumption that consumers remember product information at different levels of abstraction (Olson and Reynolds, 1983; Young and Feigen, 1975). In other words, the concrete cues trigger perceptual attributes. The evaluations of mobile service quality along perceptual attributes merge into more abstract evaluations. The more global assessment at higher levels of abstraction (e.g. perceived value, overall service quality) in turn influence behavioral intentions and actual behavior. The higher order abstractions are consequences of the anteceding evaluative process of the mobile service assessment.

The model below illustrates the causal relationships of the variables assessed in the structural equation model with the purpose of describing the usage behavior of mobile services users. This model is tested in course of the quantitative research. All the items are tested using scales from various research in that specific field. Although more constructs could be included in the model with paths well established in marketing literature, the focus of this study and a limit of complexity influenced the final model.

There is a big choice of rivalling models and the researcher has to choose between concepts. Such constructs as variety seeking or commitment could have been included. Also more antecedents of service quality would have been possible. Furthermore, more paths could have been included. One of them may be an effect of value on satisfaction. This one was not considered since the active behavior of the user like word of mouth and loyalty were in the focus of the research project. It was not the aim to explicitly measure satisfaction and the consequences of satisfaction. This would suggest another path from satisfaction to word-of-mouth. By performing favourable actions such as repete purchase (loyalty) and word-of-mouth one can assume the user is satisfied to a certain extent. The direction of the path between loyalty and word-of-mouth could also be argued about. The reasoning here is that a person first would talk about a service in a favourable way and as a consequence be a loyal customer.

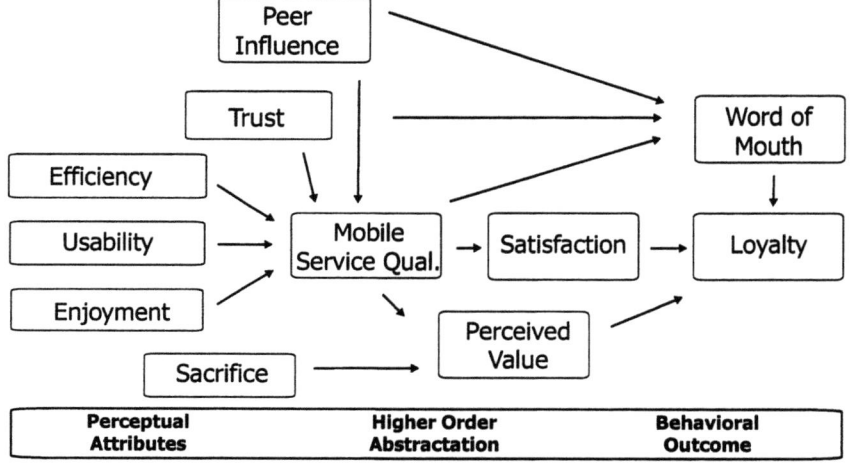

Figure 23: A Structural Model for Mobile Services User Behavior

Certainly a huge amount of data would be required including even more concepts and more paths. Therefore, one has to cut back a *total model* to a *workable* model. Such partial models as a result of theoretical self-restriction seem promising in behavioral research (Mazanec, 1982).

The model illustrates how concrete cues of a service influence higher order abstractions and beliefs to yield a certain behavioral outcome. The following table defines each construct included in the research model in order to clarify these terms.

Table 15: Definition of the Constructs Included in the Research Model

Construct	Abbrivia-tion	Definition
Service Quality	Qua	Service quality is an overall measure of the customers' perceptions on a service leading to a certain degree of satisfaction.
Loyalty	Loy	Loyalty consists of an attitude towards the service and behavioral intention to use.
Perceived Value	Val	Value is related to the consumers' trade off between what they have to give up for the consumption and their perceived value from consumption.
Word-of-Mouth	Wom	Word-of-mouth includes recommendation of the service, positive remarks and the like.
Trust	Tru	Trust relates to the correct technical functioning of a service and to the company providing the service.
Peer Influence	PI	Peer influence is the extent to which service usage is influenced by individual's peers and reference groups.
Usability	Usa	Usability relates to the extent that actual usage is free of effort.
Efficiency	Eff	Efficiency is the degree to which the service enhances the individual's performance and facilitates transactions.
Enjoyment	Enj	Enjoyment is the degree to which using a service is fun and an end in it self.
Sacrifice	Sac	Relates to what people have to give up to use a service.

Now each of the constructs is discussed in the light of the causal model exemplified above.

6.1 Perceptual Attributes

According to diffusion literature the relative advantage of a system leads to adoption (Rogers, 1995). In technology acceptance literature (Davis, Bagozzi et al., 1989; Venkatesh, Morris et al., 2003) this reasoning is picked up and included in form of two main constructs influencing the technology adoption process, perceived usefulness and perceived ease of use. Perceived usefulness is the degree to which a person believes that using a particular system enhances their performance (Davis, 1989). A system high in perceived usefulness has a positive use-performance relationship. Perceived Usefulness has a direct influence on intention to use. Often, though, people use a technology without perceiving it useful as social norms pressure them to adopt.

In service quality the efficiency of Web sites was also explored (Parasuraman, Zeithaml et al., 2005) and the relationship between efficiency and perceived quality was established. According to Hyvönen and Repo (2004) mobile services are used when they are useful for everyday life and entertaining. Wu and Wang (2005) concluded that perceived risk, cost, compatibility and perceived usefulness have a significant influence on behavioral intention to use and actual usage. In the case of the mobile service explored efficiency relates to the degree the service helps the user to save time and degree of convenience. Therefore, it is suggested that:

H1: Efficiency has a direct positive effect on the perceived service quality.

Usability is an issue well researched and often discussed concerning Web sites and computer systems (Chen and Macredie, 2005; Palmer, 2002). When it comes to interaction with mobile devices usability is even more important because of smaller screen display, more difficult navigation and input methods, and different devices (Hassanein and Head, 2003; Schmidt-Belz, A. et al., 2002). Usability relates to the degree to which a person believes that using a particular system is free of effort. This follows the definition of ease, which means freedom from difficulty or great effort. Davis (1989) claims that an application perceived to be easier to use than another is more likely to be accepted by users. Perceived ease of use is a hurdle users have to overcome for acceptance, adoption and usage of a system (Davis, 1989). Also in service quality literature a relationship between the usability of a system, in most cases Web sites, and service quality was established (Parasuraman, Zeithaml et al., 2005; Wolfinbarger and Gilly, 2003):

H2: Usability has a direct positive effect on the perceived service quality.

Enjoyment and entertainment go beyond efficiency and usability and are critical for entertainment services such as mobile games, mobile video, audio streaming, chat and flirt services (Leung and Wei, 1999). The instrumentality of this concept is entertainment itself and not the efficiency or effectiveness of being able to access mobile entertainment services ubiquitously (Pedersen, Nysveen et al., 2002). Recent approaches include emotions in ICT-adoption models (Venkatesh, 2000). This suggests including this construct in the model.

Enjoyment was included in surveys on online shopping and the intention to return to a Web site (Jarvenpaa and Todd, 1997; Koufaris, 2002). When extending the TAM with perceived enjoyment a positive relationship was found with ease of use (Moon and Kim, 2001; Venkatesh, 1999; Venkatesh, 2000) and perceived usefulness (Agarwal and Karahanna, 2000) of the system. Other surveys found an effect of enjoyment on behavior (Triandis, 1980), intention (Davis, Bagozzi et al., 1992) and usage (Teo, Lim et al., 1999):

H3: Enjoyment has a direct positive effect on perceived service quality.

6.2 Higher Order Abstractions

Service quality is an overall perception, similar to attitude in this survey. Various antecedents, which are concrete cues, influence this concept as already explained above. The literature on the relationship between quality and satisfaction is vast. Some authors state that satisfaction is an antecedent of service quality (Bitner, 1990; Bolton and Drew, 1991) while others found an effect of quality on satisfaction (Cronin and Taylor, 1992; Parasuraman, Zeithaml et al., 1985; Parasuraman, Zeithaml et al., 1988). In this research the reasoning of Cronin and Taylor (1992) is followed, therefore:
H4: Perceived service quality has a direct positive effect on satisfaction.

Like quality, value is proposed as a higher level abstraction. Value appears to be more individualistic than quality and, thus, a higher level concept. Value, as opposed to quality, involves a tradeoff between give and get components. Often quality is the only get component of value. Consumers, however, may include more higher order factors such as prestige, convenience, or trust (Holbrook, 1999).

Generally, the causal chain between quality, sacrifice and value is well established in literature (Zeithaml, 1988). Previous studies showed that perceived quality is an antecedent of perceived value. Cronin et al. (2000) found the quality – value relationship in six industries. Bolton and Drew (1991, 383) examined the service value evaluations of telephone service customers and found that service quality was a significant antecedent to service value:
H5: Perceived quality has a direct positive effect on perceived value.

To be valuable for the customer mobile services have to accomplish value propositions. They have to be personal, on time and at any place, entertaining and affordable. In the multi-channel environment where consumers are free to use various channels for information retrieval, it is essential for companies to evaluate what channels can add to existing service delivery. The main goal has to be to satisfy customer needs (Montoya-Weiss, Voss et al., 2003) and consequently create loyal customers.

A well established theory is that loyalty is earned by delivering superior value (Dodds, 1991; Dodds, Monroe et al., 1991; Reichheld, 1993; Sirohi, McLaughlin et al., 1998). A way of adding value is by offering consumers additional channels to obtain services, such as mobile data services. Therefore, it is suggested that when the consumer perceives the mobile channel as valuable the overall loyalty towards the provided service is increased.

In service literature, studies investigated the importance of value and found a significant relationship between perceived value and purchase intentions (Bolton and Drew, 1991; Woodruff, 1997; Zeithaml, 1988). In fact Parasuraman and Grewal (2000) defined value as the most important determinant of consumer purchases. Readings on mobile commerce suggest that perceived value is needed to create a critical mass to further establish mobile services (Pedersen, Leif et al., 2002; Shankar, Driscoll et al., 2003). It has to be noted that service quality, satisfaction and value are all directly related to behavioral intention (Cronin, Brady et al., 2000). Considering the nature of the constructs measured in this study a value loyalty link appears logical:
H6: Perceived value has a direct positive effect on loyalty.

Furthermore, Cronin and Taylor (1992) found that across different samples, satisfaction had a significant effect on purchase intentions. Service quality, however, did not show such an effect. Also investigating the significance tests suggests that satisfaction has a stronger and more consistent effect on loyalty (Brady, Cronin et al., 2002). Therefore, the following hypothesis is as follows:
H7: Satisfaction has a direct positive effect on loyalty.

Cost of services is currently a main hurdle to mobile service adoption. Cost certainly relates to value. Mobile content competes indirectly for a person's dispensable income across all categories. So, the cost has to be equal or less than the perceived value of the content for the user to want it. The main factors in success of mobile content is availability at a realistic price and if it really lends itself to the mobile platform. Surveys found that high cost and low connection rates lead to a low acceptance of mobile services (Erlandson and Ocklind, 1998; Schultz, 2001).

Furthermore, there is extensive literature on sacrifice, including conceptionalizations (Cronin, Brady et al., 2000). To use a product consumers sacrifice money and other resources, such as effort, time, and energy. To some consumers money is the most important factor. Anything that reduces the monetary sacrifice will increase those consumers' perceived value. Also saving time is valuable for some customers. Studies show that they are willing to spend more money if the time and effort are perceived as more costly.

The most common conceptualizations include time, money, and effort in relation to the value perceived, as sacrifice (Cronin, Brady et al., 2000). This concept is followed in this research, thus:
H8: Sacrifice has a direct negative effect on perceived value.

Subjective norms are the norms developed through external and interpersonal influence. Social norm is included as a direct determinant of behavioral intention in both, the theory of reasoned action (TRA) (Fishbein and Ajzen, 1975) and the subsequent theory of planned behavior (TPB) (Aijzen, 1991). Research on social norm has yielded different results. Mathieson (1991) for instance found no significant effect of subjective norm on intention. Taylor and Todd (1995b) on the other hand found a significant effect. Venkatesh and Davis (2000) found that social influence processes (subjective norm, voluntariness, and image) and cognitive instrumental processes (job relevance, output quality, result demonstrability, and perceived ease of use) have a significant influence on user acceptance.

Social norms are very influential in explaining the adoption and use of new media (Webster and Trevino, 1995). The subjective norm is particularly important for young users' adoption of mobile services (Leung and Wei, 1999; Ling, 2001). Young users may be more affected by external influence because their subjective norms are developing and changing. In this survey peer influence is considered. New technologies are not solely used due to their superior usability, perceived usefulness, but because of peer influence. It is suggested that peer influence has effect on the overall service quality perception and also on word-of-mouth since an innovation has to be communicated through a system at early stages of diffusion. It is suggested that peer influence has an effect of word of mouth. If the reference group is important to a person there should be more talking about a service. Also if the reference group is important the person would be more influenced by their views on the service. Thus, we propose the following hypotheses:

H9: Peer influence has a direct positive effect on word-of-mouth.

H10: Peer influence has a direct positive effect on the perceived service quality.

Trust, as a concept in research, has been examined in several social sciences, including sociology, psychology, anthropology, economics, marketing, organizational behavior and most recently information systems (Bhattacherjee, 2002). A large stream of research stresses that people can develop trust in public institutions or organizations (Morgan and Hunt, 1994), as well as persons.

The trust literature suggests that the trusting party must be vulnerable to some extent for trust to become operational (Schlenker, Helm et al., 1973). The vulnerability in m-commerce is the failing transmission of messages and security/privacy concerns. Drawing on theory from internet shopping trust can be adapted for this research. Various authors state that a lack of trust in the Web site could reduce the individual's desire to carry out transactions on the Web (Doney and Canon, 1997; Hoffman, Novak et al., 1999). Also this form of insecurity in-

volved with electronic devices holds for mobile phones. The correct functioning of the services is important for the quality perception:
H11: *Trust has a direct positive effect on the perceived service quality.*

Marketing and related literature often shows positive relationships between trust and satisfaction (Anderson and Narus, 1990; Dwyer, Schurr et al., 1987). However, it is not aim of this research to explicitly measure satisfaction but rather to investigate the effect of trust on behavior. Therefore, the relationship between trust and loyalty is explored. In services marketing Berry and Parasuraman (1991, 144) argue that *"customer company relationships require trust."* They add *"Effective services marketing depends on the management of trust because the customer typically must buy a service before experiencing it."* And in retailing Berry (1993) states that *"trust is the basis for loyalty."* Adapted to this research it needs to be mentioned that, trust is emerging as a key aspect leading to IT acceptance and can be viewed as a multi-dimensional construct combining beliefs that influence behavioral intentions (Gefen, 2002). As a psychological state, trust is distinct from, but antecedent to, behavior. The behavior related to in this context is carrying out word-of-mouth. This is the behavior selected since the social context (influence of peer influence) and communication of an innovation through a social system (Rogers, 1995) is important for the diffusion of the mobile service. Consequently, it is suggested that:
H12: *Trust has a direct positive effect on word-of-mouth.*

6.3 Behavioral Outcomes

Word-of-mouth communications received extensive attention from academics and practitioners in the last decades. Since the early 1950s, researchers have demonstrated that personal conversations and informal exchange of information influence consumers' choices and purchase decisions (Arndt, 1967; Whyte, 1983) and also can shape consumers' expectations (Zeithaml and Bitner, 1996), pre-usage attitudes (Herr, Kardes et al., 1991) and even post usage perceptions of a product or service (Bone, 1995; Burzynski and Bayer, 1977). Thus, a positive service experience and therefore a good perceived quality may lead to word-of-mouth:
H13: *Perceived service quality has a positive direct effect on word-of-mouth.*

Some surveys report the effect of word-of-mouth to be greater than print ads, personal selling and radio advertising (Engel, Blackwell et al., 1969; Feldmann and Spencer, 1965; Katz and Lazarsfeld, 1955). However, Van de Bulte and Lilien (2001) show that some of the effects might have been overrated.

Research attempts were dedicated to the antecedents and consequences of word-of-mouth with three major streams of research. First, research was conducted to find out why people proactively spread the word about products and services they have experienced. Novelty of the product, extreme satisfaction and dissatisfaction or commitment to the firm drives those behaviors. Research shows that customers with little experience in a product category seek opinions of others for product advice (Bone, 1992). Word-of-mouth is an important variable for diffusion of innovations (Mahajan, Muller et al., 1990; Rogers, 1995) even suggested as an important success factor (Bansal and Voyer, 2000; Harrison-Walker, 2001). Especially with services that are highly innovative, such as mobile data services, word-of-mouth becomes even more important (Bansal and Voyer, 2000).

For the research at hand the active part of the customer as a person that gives a dvice, talks in favour of a product, and even recomments it is important. It is believed that individuals talking in favour of a service are also more loyal to the service. Thus, the next hypothesis is as follows:

H14: Word-of-mouth has a positive direct effect on loyalty.

6.4 Moderator effects

6.4.1 Innovativeness

Individual differences can affect how individuals respond to innovations. Personal innovativeness as a construct has proven to be important to the study of individual behavior toward innovations and thus, has a long-standing tradition in innovation diffusion research (Rogers, 1995). The construct has also received attention in marketing (Flynn and Goldsmith, 1993; Goldsmith and Hofacker, 1991; Midgley and Dowling, 1978) and in information systems (Agarwal and Prasad, 1998). As attitudes and perceptions might vary among different users (innovators vs. laggards) this factor is included in the model as moderator variable. This construct can be measured adapting Goldsmith and Hofacker's (1991) product innovativeness measure to mobile services. Goldsmith has also applied this innovativeness measure for identifying Internet innovators, majority users and laggards (Goldsmith, 2001). The measure was also used by Agarwal and Prasad (1998) studying the innovativeness of Internet users. It is hypothesized that the paths are different for more and less innovative users:

H15: Innovativeness has a moderating effect on the structural relationships of the model.

6.4.2 Experience

Prior experience can be an important determinant of behavior (Ajzen and Fishbein, 1980; Bagozzi, 1981b). It has been suggested that knowledge from past behavior can help shape intention (Fishbein and Ajzen, 1975). This is due to more accessible memory (Fazio and Zanna, 1978) and the fact that events in past experience are more salient and therefore it is ensured they are accounted for in the formation of intentions (Ajzen and Fishbein, 1980).

Taylor and Todd (1995a) investigated the role of prior experience on IS usage. They used an augmented version of TAM and compared the experienced with the inexperienced user groups.

There may be differences between experienced and inexperienced users in their determinants of mobile service usage. Such differences may require alternative ways of implementation and development of new services. Therefore, it is suggested that:

H 16: Experience has a moderating effect on the structural relationships of the model.

6.4.3 Age

Age was identified as a moderator variable in research reviewed by Venkatesh et al. (2003). It is hypothesized that younger users have better perceptions with regard to mobile service use and may show different behavioral patterns. Age has shown to have an effect in the technology acceptance context (Venkatesh and Morris, 2000). In a job context age has shown to be associated with processing stimuli and allocating attention to the job (Plude and Hoyer, 1985). In this research age is coded as a continuous variable, consistent with prior research (Venkatesh and Morris, 2000). It is suggested that younger users have different perceptions and behave differently than older users:

H17: Age has a moderating effect on the structural relationships of the model.

7 Operationalization of the Constructs

The questionnaire contained multiple measures of each concept of the model. Regarding instrument construction, the items used to operationalize the constructs of each investigated variable were mainly adopted from relevant previous research. Validation and wording changes were carried out where necessary. The chosen sets of questions were translated from English into German and back translated from two different linguists. Then, in order to secure semantic consistency, a native speaker compared the original English version with the back translation. Adjustments were necessary for two questions.

Usability was measured using four items developed from adapting the original items of Davis (1989) and Davis, Bagozzi et al. (1989). A similar procedure is found in Taylor and Todd (1995b) and in Battacherjee (2000). Similar measures have been used in comparable surveys of technology acceptance and IS quality assessment (Barnes and Vidgen, 2002; Moon and Kim, 2001; Moore and Benbasat, 1991). The term system was replaced by mobile service and the back translation lead to semantically equivalent versions.

Perceived service quality, satisfaction, and perceived value are overall measures in this study. Because of the problem of generalizing product-specific attributes there has been a need for higher-level abstract dimensions of quality (Zeithaml, 1988) to make the concept of quality applicable in a general research setting. Overall quality contrasts to a composite of attributes or belief-based evaluation often termed quality performance (Cronin, Brady et al., 2000; Parasuraman, Zeithaml et al., 1988). The global assessment of quality has strong generalizability, whereas the attribute-based approach can identify specific strengths and weaknesses of a good or service. To overcome the problem of diverse beliefs about product attributes some researchers view perceived quality as an overall evaluation (Olshavsky, 1985) similar to attitude (Spreng and Mackoy, 1996; Zeithaml, 1988). This line of reasoning is followed in this survey.

Quality and value were operationalized by adopting the measures of Cronin, Brady et al. (1997; 2000). Analogous to adjusting these concepts to the mobile service setting perceived sacrifice was adjusted from Cronin, Brady and Hult's (2000) survey. They refined the previously developed measure (Cronin, Brady et al., 1997) and reused three of the originally nine items in their year 2000 study. These three are also used and adapted in this research.

Efficiency was measured using three items. One was adapted from Parasuraman, Zeithaml et al. (2005) and the other two from Childers, Carr et al. (2001).

Enjoyment was measured using three items. Two were adapted from Davis, Bagozzi et al.'s (1992) work on intrinsic motivation for the use of computers. The third one was newly developed for this survey.

The items used by Battacherjee (2000) were used to measure peer influence. Taylor and Todd (1995b) use similar items for their model in information technology usage. Three measures are adapted for the setting of m-services.

Loyalty and word-of-mouth are often operationalized as one construct – loyalty. In this survey the two are regarded as two different concepts. However, the items used for word-of-mouth were grouped as loyalty in other surveys. Nevertheless, we want to refer to them exclusively as an individual construct. Similar items were already used by several authors (Parasuraman, Zeithaml et al., 1994b; Parasuraman, Zeithaml et al., 2005). Loyalty consists of items that measure the willingness to use a service again. Besides from this behavioral component it comprises an attitudinal one (Agarwal and Karahanna, 2000; Ajzen and Fishbein, 1980; Cronin, Brady et al., 2000; Parasuraman, Zeithaml et al., 2005). Also, two new items were used to measure loyalty. They include some harder loyalty criterion with regard to a competitor offering the same service at a cheaper price.

Trust consists of two components, trust in the system availability and functioning, and trust in the company protecting the customers' privacy and secure payment. The first one is covered by using items of the questions developed by Parasuraman, Zeithaml et al. (2005). The second part is covered drawing on the items developed for Wolfinbarger and Gilly's (2003) eTailQ. Adjustments are made with regard to the wording, as opposed to Web sites subject of interest in this study are mobile services.

The construct innovativeness was measured adapting the scale used by Agarwal and Prasad (1998) studying the innovativeness of Internet users. Two of these items were applied except for replacing the term *Internet* with the term *mobile service*. For their personal innovativeness scale Agarwal and Prasad (1998) relied on Goldsmith and Hofackers' (1991) conceptual definition and of personal innovativeness as well as on their recommendations. Goldsmith also applied this innovativeness measure for identifying Internet innovators, majority users, and laggards (Goldsmith, 2001).

Age, the second moderator was determined by asking the respondents how old they were. The third moderator, experience, uses two statements indicating the duration since when mobile services have been used and in addition how often. A third question regarding the degree to which the person knows the service should indicate the users' familiarity with the service.

Scales were developed using a multi-stage procedure. First, existing scales were reviewed for their fit with the conceptual definitions of the dimensions of the model. Then, an initial set of items was constructed drawing upon prior work and the underlying conceptualization for this survey. The scales were pre-tested using a sample of 42. The result of the pre-test led to further refinement and to establish convergent and discriminant validity. All items were measured on a four-point Likert-type scale with anchors ranging from *strongly agree* to *strongly disagree* and from *high* to *low*. To ensure the desired balance and randomness in the questionnaire the items were randomly sequenced to reduce potential ceiling (or floor) effects, which includes monotonous responses to the measures of a particular construct. Effort has been made to ensure each statement captured the intended meaning of a specific sub-dimension of the constructs.

Table 16 shows the final constructs, items and sources used in the questionnaire and the final questionnaire can be found in the appendix. The scales are referred to as ordered categorical since the data can not be assumed to be interval scaled but ordinal at best. It is also assumed that the data will not be normal distributed since only four answer categories hardly yield normally distributed data.

Table 16: Operationalization of the Constructs

Construct	Number of Items	Scale	Operationalization	Adapted From
Service quality	4	Ordered categorical	1=agree – 4=disagree	(Cronin, Brady et al., 2000)
Satisfaction	1	Ordered categorical	1=agree – 4=disagree	
Peer influence	3	Ordered categorical	1=agree – 4=disagree	(Bahattacherjee, 2000; Taylor and Todd, 1995a)
Usability	4	Ordered categorical	1=agree – 4=disagree	(Davis, 1989; Davis, Bagozzi et al., 1989)
Efficiency	3	Ordered categorical	1=very – 4=not at all	(Childers, Carr et al., 2001; Parasuraman, Zeithaml et al., 2005)
Enjoyment	3 (1 New)	Ordered categorical	1=very – 4=not at all	(Davis, Bagozzi et al., 1992)
Word-of-mouth	3	Ordered categorical	1=agree – 4=disagree	(Parasuraman, Zeithaml et al., 2005)
Loyalty	4 (2 New)	Ordered categorical	1=agree – 4=disagree	(Parasuraman, Zeithaml et al., 1994b) (Cronin, Brady et al., 2000) (Ajzen and Fishbein,

				1980) in (Agarwal and Karahanna, 2000)
Perceived value	3	Ordered categorical	1=high – 4=low	(Cronin, Brady et al., 2000) (Parasuraman, Zeithaml et al., 2005)
Sacrifice	3	Ordered categorical	1=high – 4=low	(Cronin, Brady et al., 2000)
Trust	2+2	Ordered categorical	1=agree – 4=disagree	(Parasuraman, Zeithaml et al., 2005) (Wolfinbarger and Gilly, 2003)
Moderators: Innovativeness Age Experience	2 1 3	Ordered categorical; Nominal; Nominal and Categorical;	1=agree – 4=disagree yes/no Often, seldom...etc. 1=agree – 4=disagree	(Agarwal and Prasad, 1998)

8 Data Collection

Quantitative research is often associated with surveys or experiments and is still considered the foundation of the research industry for collecting marketing data.

This form of research emphasizes standard questions and predetermined response options in questionnaires. The number of respondents is very high too. Quantitative research methods are more directly related to descriptive and causal research designs than to exploratory. The information needs are precise and well defined. Success lies mainly in designing and administering the survey instrument correctly. In quantitative research practices, researchers have to know about construct development, scale measurement, questionnaire design, sampling, and statistical data analysis skills – issues dealt with in the following chapters.

8.1 Questionnaire Design

The questionnaire was designed following general rules for questionnaire design. Details are in literature included in the development process but are not repeted here, thus, the reader may want to refer to the references (Bagozzi, 1994; Malhotra, 1991; Sudman and Bradburn, 1982). The type of questions used was closed ones to facilitate analysis with appropriate software tools. The literature on asking questions is extensive and the reader can go into more detail with one of the main textbooks in this area (Green, Tull et al., 1988, 268; Kumar, Aaker et al., 2002, 238; Malhotra, Hall et al., 2002, 413).

The survey was conducted online. Since, no accompanying letter was possible the first page of this Web questionnaire served this purpose. It provided information on the institution carrying out the survey, the reason for approaching the potential respondent, the respondents' advantage through completing the questionnaire (incentives), the aim of the research, the duration of completing the questionnaire, and finally, to thank the respondent in advance for the time consumed completing the questionnaire (Paxson, 1995). All these tools were used for increasing the response rate. There are tow main types of incentives, a small incentive for everybody or a bigger one that is given away in the form of a lottery among the respondents. In this survey the lottery among the first 100 respondents is chosen, Grossnickle and Raskin (2001) also consider this method more efficient and it also motivates to complete the questionnaire immediately to be included in the lottery. A negative effect of the incentive could be a bias of respondents only participating due to the prize.

8.2 Sampling and Sample Size

Sampling is defined as: *"The selection of a small number of elements from a larger defined target group of elements and expecting that the information gathered from the small group will allow judgements to be made about the larger group." (Hair, Busch et al., 2000)*

The sample should allow conclusions regarding the population. Thus, the sample should provide a representative distribution of the attributes of the whole population (Berekoven, Eckert et al., 2001). There are different sampling techniques, among those non-probability sampling (convenience sampling, judgmental sampling, quota sampling and snowball sampling) and probability sampling (simple random sampling, systematic sampling, stratified sampling, cluster sampling, two/one stage sampling, proportionate/disproportionate sampling) (Malhotra, Hall et al., 2002). Each of these methods bears some advantages and disadvantages and, depending on the information sought in the survey, the appropriate sampling technique has to be chosen. The characteristic sought in this survey – mobile service usage - is rather rare in the population. Thus, the target population has to be well chosen. In the research at hand the target population is users of mobile data services in general and m-parking in particular. Thus, they were targeted via a web-site where they have to register and pay for m-parking. This type of sampling can be referred to as judgmental sampling, based on the judgment and expertise of the researcher that the elements are representative for the population of interest (Malhotra, Hall et al., 2002).

Another issue is sample size, which is particularly important for path analysis. The identification of an appropriate sample size is rather difficult. The complexity of the model can be measured by the parameters that need to be estimated. Bollen (1989, 298) for instance, remarks cryptically:

"...Though I know of hard and fast rule, a useful suggestion is to have at least several cases per free parameter..."

Bagozzi (1982, 380) and Bentler and Chou (1987) are more precise and argue that the sample size has to be at least the parameters estimated times five. According to Sachs (1982) sample size should be the parameters to be estimated times six.

Rules of thumb for sample size are in several sources (Fornell, 1983). Some show that small sample sizes are not compatible with maximum likelihood estimation of covariance structure models. The sample size varies depending on the size of the model and the covariance structure. Boomsma (1982; 1985) recommends a sample size of at least 100 better 400, Anderson and Gerbing (1984) 150, Bagozzi (1981a) found that maximum likelihood estimation in covariance structure analysis requires the sample size minus the number of parameters to be greater than 50.

Relevant for the survey at hand is a recommendation by Muthén and Muthén (1998) of a sample size bigger than 200 when using the WLMSV-estimator. As this estimator will be used to carry out the analysis 200 respondents are the lower limit for this study. Therefore, a final sample size of 502 m-parking users for the model test is adequate.

8.3 Pretest

In order to identify and eliminate potential problems with the questionnaire a small sample of respondents pretested it.

Pretesting answers two broad questions: i) whether or not *"good"* questions are asked ii) whether or not the questionnaire flows smoothly and the question sequence is logical (Green, Tull et al., 1988, 185). However, questionnaire validity is not secured by a pretest.

Carrying out a pretest, strongly recommended in empirical research (Babbie, 1998; Green, Tull et al., 1988; Grossnickle and Raskin, 2001), is done in this research project.

Grossnickle and Raskin (2001) suggest that for online surveys the technology side also has to be thoroughly tested. This is done in this survey with regard to download times and transfer of data from the questionnaire into the data base, etc.

The pretest was carried out in May 2005. The questionnaire was tested online by 42 respondents who were representative for the survey sample. The participants of the pretest made comments regarding the questionnaire and returned the results within two days. The author adjustmented based on the respondents' suggestions arrive with the final questionnaire.

8.4 The Field Phase

A link to the questionnaire was provided on important Web sites for m-parking. The first link was on the home page for m-parking (www.m-parking.at) where users regularly upload their account and search for information. The second link was on the Viennese government site (www.wien.at) providing citizens with information about the municipality. When the survey was carried out a special marketing campaign for m-parking was done by the city of Vienna and, as a consequence, attracted visitors on the Web site providing information and the link to this survey. The questionnaire was accessible online from June 22 2005 to August 8 2005.

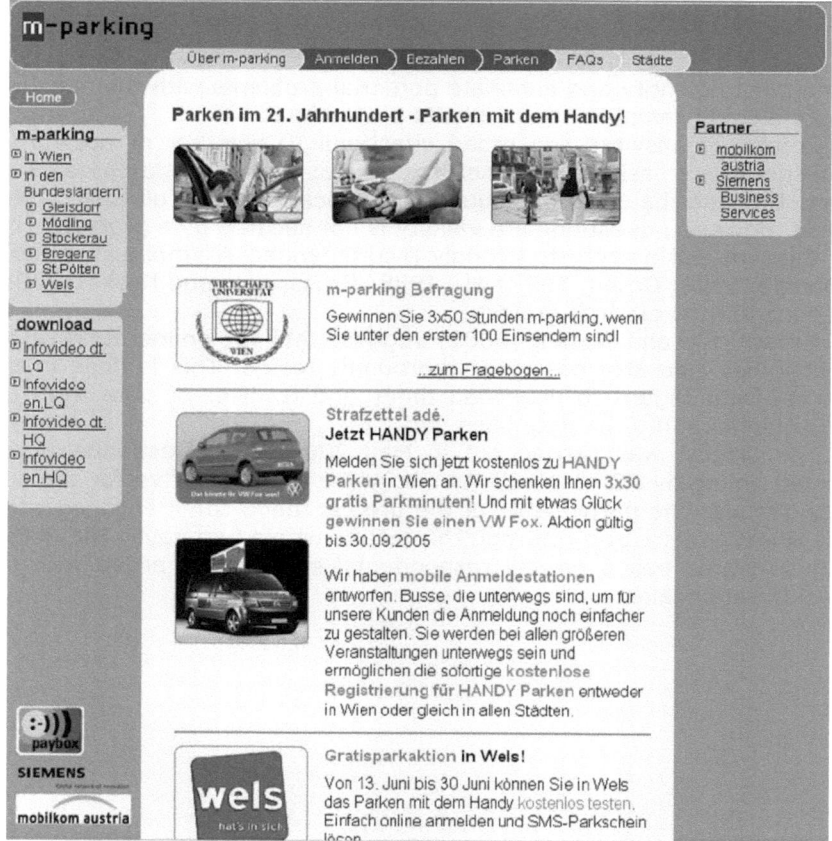

Figure 24: Link Provided on the Official m-parking Web Site

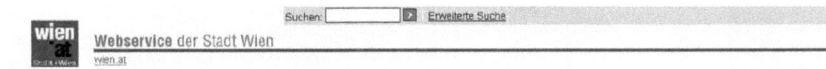

m-parking - Anmelden leicht gemacht

Täglich werden über 7.000 Parkscheine mit dem Handy gelöst, und die Tendenz steigt. Bereits 80.000 Wienerinnen und Wiener nutzen den bequemen Handy-Dienst, jeden Monat kommen durchschnittlich 1.500 neue Anmeldungen dazu. Die einmalige Registrierung erfolgt über SMS oder online über www.m-parking.at. Betrieben wird m-parking in der Bundeshauptstadt von der Stadt Wien in Kooperation mit Siemens Business Services und der mobilkom austria. Alles über die einfache Anmeldung zum m-parking erfährt man auch in den so genannten Anmelde-Bussen.

Die Wirtschafts-Universität Wien führt eine Studie zum Thema m-parking durch - machen Sie mit und nutzen Sie Ihre Chance, 3x50 Stunden m-parking zu gewinnen.
Online-Fragebogen

Standorte der m-parking-Anmelde-Busse

Hinweis: Terminänderungen vorbehalten! Veranstaltungen können kurzfristig - z.B. wegen Schlechtwetter - verlegt werden.

Datum	1. Anmelde-Bus	2. Anmelde-Bus	Event
Freitag, 1. Juli	Praterstern U-Bahnstation, 2. Bezirk		
Samstag, 2. Juli	Praterstern U-Bahnstation, 2. Bezirk	Freihausviertel-Fest auf der Wieden, 4. Bezirk	Regenbogenparade - Ringstraße, 1. Bezirk
Sonntag, 3. Juli	Gänsehäufel (Bad), 22. Bezirk	Schafbergbad, 19. Bezirk	
Montag, 4. Juli	Am Hof, 1. Bezirk	Schottentor, 1. Bezirk	
Dienstag, 5. Juli	U-Bahnstation Heiligenstadt, 19. Bezirk	U-Bahnstation Spittelau, 9. Bezirk	
Mittwoch, 6. Juli	Meidlinger Hauptstraße 57-59, 12. Bezirk	Meidlinger Platzl, 12. Bezirk	
Donnerstag, 7. Juli	Mariahilfer Straße - Europaplatz, 15. Bezirk	Mariahilfer Straße - Kirche, 6. Bezirk	

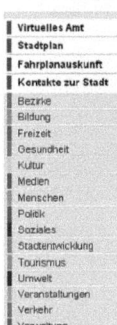

Figure 25: Link Provided on wien.at

9 EMPIRICAL TEST AND ANALYSES

9.1 Response Rate

Altogether, the questionnaire was accessed 1598 times and fully completed by 1248 respondents. Five cases were deleted as they gave the same answer throughout the questionnaire. Four who gave unreliable details with regard to their demographic profile (e.g. 81 years old, self employed, gender male, female first name...). Furthermore, two were deleted since they reported to use the m-parking service in a city where it is unavailable. Two more were deleted since they filled out the questionnaire twice. Four were excluded since they reported they never used mobile services but later on reported they used m-parking which leads to the assumption they ticked answers at random. Finally 1231 cases remained in the data set for further analysis. 41% of the respondents were real m-parking users and can be used in the model and hypotheses tests. A skip logic included in the questionnaire guided the non users directly to the questions to obtain the demographic data. This demographic profile of non users allows a comparison with the users.

At the weekend the response was lowest; on average 26 respondents filled out the questionnaire per day with a maximum of 70 respondents on one Tuesday. Generally the peek of responses was in the middle of the week, which is shown by the following graph. On the week ends the response rate flattened out.

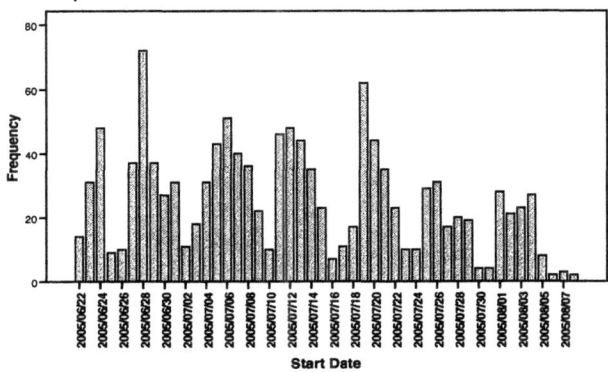

Figure 26: Frequencies for Filling out the Questionnaire During the Field Phase

With regard to duration for filling out the questionnaire the average was five minutes. It took some outliers up to two hours to complete the questionnaire and some were as fast as 44 seconds. These were also excluded. The fastest ones were m-parking non-users with a considerably shorter questionnaire only asking for usage habits of other services and demographic details.

9.2 Sample Structure

The following section gives details on the m-parking users and non-users in the sample.

Table 17: Profile of the Whole Sample

Gender:		**Income:**	
Male	57.0%	0-1000 Euro	21.1%
Female	40.7%	1,001 – 2,000 Euro	44.0%
Missing	2.3%	2,001 – 3,000 Euro	19.3%
Age:		more than 3,001 Euro	7.4%
Average	34.44	*Missing*	8.1%
Maximum	69	**Occupation:**	
Minimum	16	Pensioner	3.1%
Percentile 25	26.00	In education	0.2%
Percentile 50	32.00	Self employed	14.1%
Percentile 75	40.00	Employee	61.9%
Missing	2.3%	Student	15.2%
Highest Completed Education:		Unemployed	3.1%
Elementary School	2.7%	*Missing*	2.4%
Secondary Education	21.4%		
A-Levels	39.3%		
University	34.1%		
Missing	2.4%		

9.2.1 Profile of m-parking Users and Non-Users

A total of 502 (40.8%) users fully completed the questionnaire. The following table summarizes the characteristics of this sub-sample.

Table 18: Characteristics of the Sub-Sample "Users"

Gender:			**Income:**	
Male	60.6%		0-1000 Euro	12.0%
Female	34.1%		1,001 – 2,000 Euro	24.6%
Missing	5.4%		2,001 – 3,000 Euro	22.5%
Age:			More than 3,001 Euro	9.6%
Average	36.03		*Missing*	13.3%
Maximum	75		**Occupation:**	
Minimum	19		Pensioner	3.2%
Percentile 25	27.00		In education	0.2%
Percentile 50	35.00		Self employed	16.9%
Percentile 75	42.00		Employee	64.4%
Missing	5.5%		Student	7.9%
Highest Completed Education:			Unemployed	1.8%
Elementary School	2.2%		*Missing*	5.6%
Secondary Education	21.9%			
A-Levels	37.8%			
University	32.5%			
Missing	5.6%			

The following table provides a profile of the 729 (59.2%) non-users.

Table 19: Profile of the Sub-Sample M-parking Non-Users

Gender:			**Income:**	
Male	54.6%		0-1000 Euro	27.4%
Female	45.3%		1,001 – 2,000 Euro	54.0%
Missing	0.1%		2,001 – 3,000 Euro	17.1%
Age:			More than 3,001 Euro	5.9%
Average	33.43		*Missing*	4.5%
Maximum	69		**Occupation:**	
Minimum	16		Pensioner	3.0%
Percentile 25	26.00		In education	0.3%
Percentile 50	30.00		Self employed	12.1%
Percentile 75	39.00		Employee	61.6%
Missing	0.1%		Student	20.2%
Highest Completed Education:			Unemployed	2.6%
Elementary School	3.0%		*Missing*	0.3%
Secondary Education	21.1%			
A-Levels	40.3%			
University	35.3%			
Missing	0.3%			

The biggest difference between the two groups is income. Non-users tend to earn less than m-parking users.

9.2.2 M-Services Usage Behavior

The respondents are analyzed with regard to their mobile services usage behavior. First of all it is interesting which mobile phone provider the customers use. Some providers follow a low cost strategy, hardly providing any additional mobile services (Yess, telering) while others are market leaders in providing mobile services (A1). The following table presents the provider distribution of all users and the m-parking users and non-users separately in percent. The questionnaire allowed for multi-response since some people have more than one mobile phone. The results indicate that the biggest share of users are customers of A1 and one.

Table 20: Provider Distribution of the Respondents

Provider	All respondents	Users	Non-Users
A1	29.3 %	33.7 %	26.4 %
t-mobile	23.1 %	19.3 %	25.6 %
One	25.3 %	31.6 %	21.0 %
telering	17.1 %	11.5%	20.9 %
Drei	4 %	3.5 %	4.3 %
Yess	1.2 %	0.4 %	1.7 %

The following table shows which mobile services are used by the respondents.

Table 21: Mobile Services Already Used by the Respondents

M-Service	All respondents	m-parking users	Non-Users
Information Service	17.6 %	13.5 %	23.7 %
Financial Service	6.7 %	6.4 %	7.2 %
Ticketing	10.1 %	9.0 %	11.6 %
Advertising	3.0 %	2.4 %	3.9 %
LBS	5.6 %	5.2 %	6.2 %
Entertainment	18.2 %	12.4 %	26.6 %
Bus/Tram/Underground Ticket	6.3 %	7.1 %	5.1 %
Shopping	3.1 %	3.3 %	2.8 %
Lottery	4.4 %	4.7 %	3.9 %
m-parking	19.8 %	31.4 %	3.1 %
Other	5.3%	4.8 %	6.0 %

The results show that, besides from m-parking, users of this service also have experience with information services and entertainment services. Among the non-users the experience with information services and entertainment services ranks even higher with 23% and 26% respectively. This hints that the non m-parking users might be more fun-oriented.

The following pie charts show the distribution of services m-parking users and non-users would like to use in the future. Among the non-users 24% would like to use m-parking in the future. In summary, both groups have a strong interest in using tickets for public transport (10% for the whole sample), information services (17% for the whole sample) and entertainment services (18% for the whole sample).

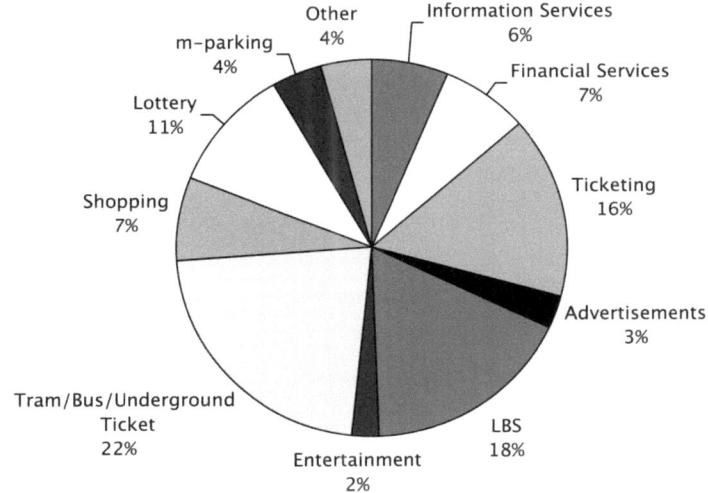

Figure 27: Services m-parking Users Would Like to Use in the Future

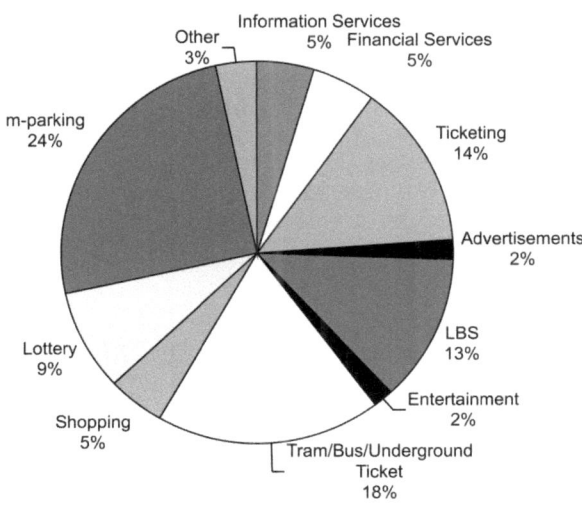

Figure 28: Services m-parking Non-Users Would Like to Use in the Future

123

Regarding the frequency of m-services usage, there is a bigger difference between the two groups. The following cross tabulation shows the frequency of m-service usage between m-parking users and non-users.

Table 22: Cross Tabulation of Frequency of Mobile Service Use and Usage of M-parking

			m-parking		Total
			Yes	No	
Frequency	More than once a week	% of frequency	75.4%	24.6%	100.0%
		% of usage m-parking	42.6%	9.6%	23.1%
	Once a week	% of frequency	62.4%	37.6%	100.0%
		% of usage m-parking	27.1%	11.2%	17.7%
	Once a month	% of frequency	37.2%	62.8%	100.0%
		% of usage m-parking	25.1%	29.2%	27.5%
	Once a year	% of frequency	16.4%	83.6%	100.0%
		% of usage m-parking	5.2%	18.2%	12.9%
	Never	% of frequency		100.0%	100.0%
		% of usage m-parking		31.7%	18.8%
Total		% of frequency	40.8%	59.2%	100.0%
		% of usage m-parking	100.0%	100.0%	100.0%

Table 22 illustrates the frequencies per row and per column. The results indicate that m-parking users use mobile services more often than non-users. This suggests that users of m-parking services are generally heavier m-services users. The following table shows if the m-parking users are just the heavier users or if they also have been using mobile services for a longer period of time.

Table 23: Cross Tabulation of Duration of Mobile Service Use and Usage of M-parking

			m-parking		Total
			Yes	No	
Usage since	>3 years	% of usage since	41.8%	58.2%	100.0%
		% of usage m-parking	23.3%	22.4%	22.7%
	2 years	% of usage since	54.3%	45.7%	100.0%
		% of usage m-parking	34.3%	19.9%	25.8%
	1 years	% of usage since	47.6%	52.4%	100.0%
		% of usage m-parking	26.1%	19.8%	22.3%
	0,5 years	% of usage since	63.6%	36.4%	100.0%
		% of usage m-parking	16.3%	6.4%	10.5%
	Never	% of usage since		100.0%	100.0%
		% of usage m-parking		31.6%	18.7%
Total		% of usage since	40.8%	59.2%	100.0%
		% of usage m-parking	100.0%	100.0%	100.0%

The second cross tabulation in above table points at the fact that there might be sub-groups among the non m-parking users. Some of them (31%) never used any mobile services and the second rather big one has been using mobile services for more than three years (22%) or two years (19%). Among the m-parking users are more long-term m-services customers. Nearly six out of ten from this group have been using m-services since two years or more.

9.2.3 M-Parking Usage

This section gives details on the m-parking usage. Most m-parking users got aware of the service through friends (22%), papers (18%), Internet (16.5%), radio (9%), and TV (8%).

M-parking is most frequently used in Vienna (90%) even though it exists in several other cities. These are Bludenz (.2%), Bregenz (.4%), Kitzbühel (.2%), Krems (1.3%), Mödling (3%), St. Pölten (.7%), Stockerau (.6%), Tulln (.6%), and Wels (2.4%). In Amstetten, Kitzbühel, and Stockerau m-parking is also offered but no respondents from there completed the questionnaire. The low respondent figures may result from Vienna being by far the biggest city among the ones providing m-parking. There the number of registered users is above 50,000. In the other cities the service just started to operate and not even a couple of hundred may have registered yet. Another indicator is that Vienna is also the biggest city in Austria with most parking lots available. In addition the usual form of paying the parking fee in Vienna is inconvenient for drivers. This is also reflected by the results of the question about the main reason for using m-parking. Most users registered as they disliked by the inconvenient way of paying for the parking fees.

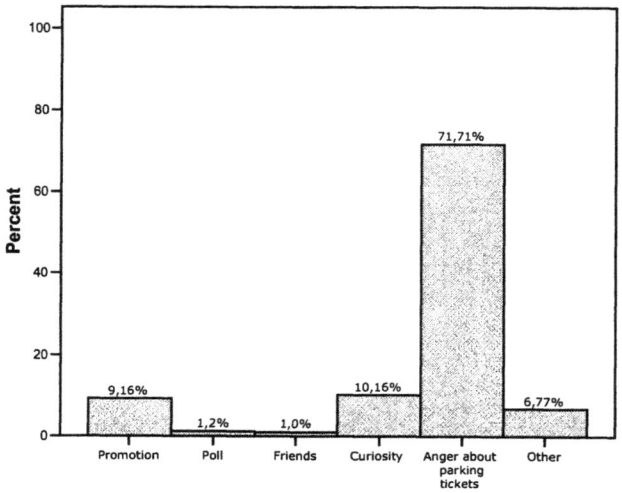

Figure 29: Reasons to Register for M-parking

The majority, 75%, registered for m-parking via the Internet and nearly six out of ten registered within a week after becoming aware of the m-parking service.

9.2.4 Data Structure

For the model test, and choice of the appropriate estimator, the relevant variables must be tested for normal distribution. This was carried out applying the Kolmogorov-Smirnov Test. The data is not normal distributed. Most of the items were skew which was graphically shown by plots including the frequencies of observations and a normal curve.

9.3 Test Results

9.3.1 Test of the Hypothesized Model

As a first step the full hypothesized model is tested. As stated above 502 m-parking users are included in the analysis. The analysis is carried out using M-Plus applying the Muthén estimator for categorical data.

There is no missing data as skipping a question was not permitted in the questionnaire. Only data regarded as rather private such as income and demographic details were allowed to be skipped. Thus, it is not necessary to do the analysis with and without missing values.

The following figure shows the hypothesized model and measurement model. It also includes the estimates for the full model. The figures in parentheses are not significant at the p=.05 level. All coefficients depicted are standardized.

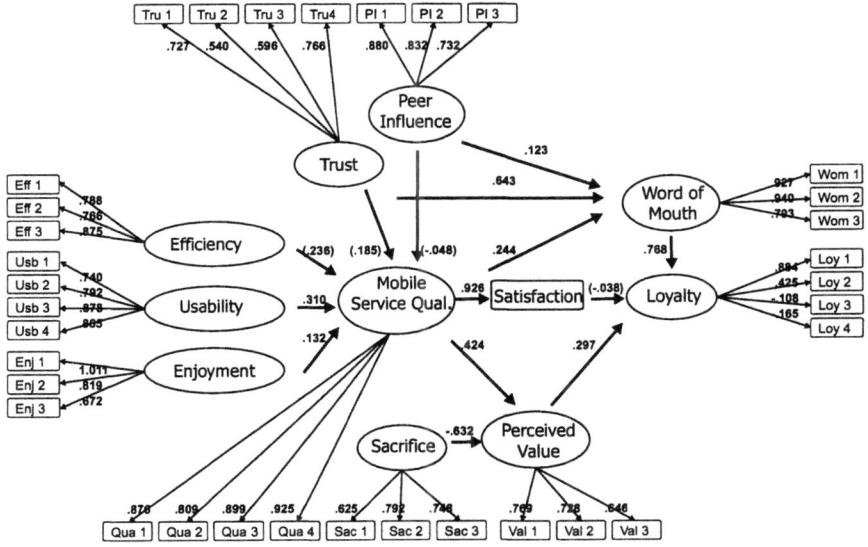

Figure 30: Full Research Model

Table 24: Fit Indices for the Full Research Model

Fit Index	Obtained Level
CFI	.826
TLI	.948
RMSEA	.104
SRMR	.072
WRMR	1.758

Mplus reports the fit measures given in Table 25. Of the two incremental fit measures, TLI is above the required level whereas CFI remains below. RMSEA, an absolute fit measure is above the required .08 level and indicates that the structural/measurement model does not quite predict the covariance matrix. SRMR is close to the required level of .06. In summary, there is a satisfying model fit.

The results support most of the hypothesized relationships. Out of the 14 hypotheses four were insignificant. Thus, the interpretation of the paths has to be carried out carefully.

According to the findings the strongest determinant of mobile service quality is usability. This is in line with research that suggests that usability has a positive effect on perceived quality. An other positive, significant path can be found between enjoyment and perceived quality. All the other hypothesized constructs (trust, peer influence, and efficiency) did not show significant results. Thus, these paths will not be interpreted here.

The other two hypotheses involving trust and peer influence could be supported. Trust shows a particularly strong effect on word-of-mouth with a coefficient of .64. Also the hypothesized path between peer influence and word-of-mouth could be supported with .123.

The three hypotheses involving the effect of mobile service quality on satisfaction, on perceived value, and on word-of-mouth could all be supported with coefficients of .926, .424, and .244 respectively. The effect on satisfaction, the strongest one, is also well established in literature. This relationship could be supported. Also the relationship between perceived quality and value has received a lot of attention in literature and is also supported in this study. Continuing the discussion with regard to the value construct, sacrifice showed, as hypothesized, a strong negative effect. It is evident that the cost, time, and effort involved in using m-parking have a negative impact on the perceived value. The quality of the service, on the other hand, has a strong positive one. Perceived value has a positive effect on loyalty (.297) while the effect of satisfaction on loyalty is low and not significant. The third hypothesized effect on loyalty, the one of word-of-mouth, is very strong. This could be caused by various reasons, for instance the measurement model, which is rather weak for loyalty and has to be tested more thoroughly.

Also, the high coefficient between word-of-mouth and loyalty could be caused by a lack of discriminant validity. The issues involving the measurement instrument are investigated in more detail in the following chapter. An analysis of the measurement model may detect weaknesses that influence the constructs and consequently the structural model. Some authors suggest the separate investigation of the structural and the measurement model to get a more comprehensive view (Anderson and Gerbing, 1988).

The following table shows the hypotheses, the critical ratio, and gives evidence if the hypothesis was supported or not. Significance of the construct is calculated in Mplus by providing the critical ratio. This is the unstandarized coefficient divided by the standard error. Indicator reliability shows the variance explained by an item. It should not be between 1.69 and -1.69 the values higher and lower are significant at p=.05.

Table 25: Summary of the Findings of the Hypotheses Test

Hypotheses	Path Coefficient	Critical Ratio	Hypotheses Supported
H1 Eff -> Qua	.236	1.108	No
H2 Usa -> Qua	.310	2.447	Yes
H3 Enj -> Qua	.132	2.741	Yes
H4 Val -> Loy	.297	4.918	Yes
H5 Sac -> Val	-.632	-8.182	Yes
H6 Pi -> Wom	.123	2.516	Yes
H7 Pi -> Qua	-.048	-1.051	No
H8 Tru -> Qua	.185	1.156	No
H9 Tru -> Wom	.643	8.516	Yes
H10 Wom -> Loy	.768	15.100	Yes
H11 Sat -> Loy	-.038	-1.160	No
H12 Qua -> Sat	.926	43.343	Yes
H13 Qua -> Wom	.244	3.762	Yes
H14 Qua -> Val	.424	7.518	Yes

9.3.1.1 Test of the Measurement Model

Most scales in this survey stem from the scientific literature. However, in some cases new variables were used. When estimating measurement and structural models simultaneously this might suffer from interpretational confounding (Burt, 1973). Interpretational confounding occurs *"as the assignment of empirical meaning to an unobserved variable which is other than the meaning assigned to it by an individual a priori to estimating unknown parameters"*. (Burt, 1976, 4) Above that, according to Anderson and Gerbing (1988) this empirically defined meaning may change considerably, depending on the specification of free and constrained parameters for the structural model. Thus, they recommend a two step approach assessing the structural model and the measurement model independently. However, this is an exploratory exercise. Generally a model should only be rejected considering measurement and structural model since both are part of the developed theory.

The causal model consists of 34 items with all of them categorical. According to Magidson (1982) categorical data are rarely normally distributed. Thus, use of statistical techniques assuming normal distribution is not feasible. In the causal model not the means but the thresholds are used. Thresholds are estimated for binary and ordered categorical observed variables in Mplus. The sign of a threshold is the opposite of the sign or a mean or intercept of the same fariable. For instance, with a binary dependent variable, a threshold of -0.5 is the same as an intercept of .5. A threshold is the underlying distribution Y* of a latent variable where it becomes y=1 (otherwise y=0). If data are binary there is one threshold, when data are polytomous with x categories there are x-1 thresholds.

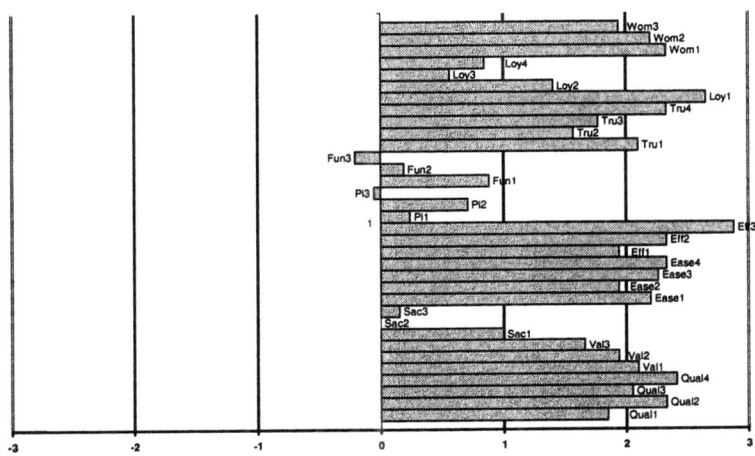

Figure 31: Thresholds of the Categorical Variables

Figure 31 shows the thresholds Mplus estimates for the distribution of Y* of the indicators. Similar to the frequencies the thresholds of some variables are close to the extreme points of the distribution. This could lead to problems with regard to the estimator.

In Mplus only eight characters are allowed to name the variables. These short names are used in the following table. It shows the items' means, standard deviations, Cronbach's alphas, factor loadings, R^2, and critical ratio. The means and standard deviation again, as reported earlier, show that data are not normal distributed.

Table 26: Item Mean, Standard Deviation, Cronbach's Alpha, and Factor Loading

Item	Mean	Standard Deviation	Cronbach's Alpha	Factor Loading	Critical Ratio	R^2
Qual1	1.75	.776	.867	.888	0.000	.789
Qual2	1.44	.615		.816	27.531	.666
Qual3	1.67	.725		.914	40.891	.835
Qual4	1.42	.576		.925	37.050	.856
Sat	1.47	.624				
Val1	1.57	.719	.645	.763	0.000	.583
Val2	1.86	.747		.727	17.165	.528
Val3	1.68	.800		.643	14.448	.413
Sac1	2.68	.835	.654	.646	0.000	.418
Sac2	3.36	.760		.825	12.322	.681

Item						
Sac3	3.28	.767		.775	11.515	.600
Usab1	1.52	.682	.817	.739	0.000	.546
Usab2	1.63	.708		.792	21.117	.628
Usab3	1.57	.673		.876	25.659	.767
Usab4	1.47	.643		.866	22.645	.750
Eff1	1.49	.703	.747	.789	0.000	.623
Eff2	1.39	.621		.768	21.585	.590
Eff3	1.35	.577		.873	24.879	.761
Pi1	2.90	1.097	.777	.882	0.000	.778
Pi2	2.37	1.113		.831	14.740	.691
Pi3	3.14	1.058		.733	12.776	.538
Enj1	2.33	1.060	.769	1.007	0.000	
Enj2	3.06	.991		.821	14.489	.674
Enj3	3.35	.892		.673	12.919	.452
Tru1	1.46	.679	.667	.779	0.000	.607
Tru2	1.62	.850		.578	11.115	.334
Tru3	1.81	.756		.640	12.683	.410
Tru4	1.59	.650		.831	17.232	.691
Loy1	1.24	.492	.402	.877	0.000	.769
Loy2	2.19	.872		.431	9.422	.186
Loy3	2.80	1.006		-.112	-2.120	.013
Loy4	2.36	1.101		-.170	-3.123	.029
Wom1	1.44	.619	.848	.923	0.000	.851
Wom2	1.42	.630		.939	36.480	.881
Wom3	1.68	.760		.798	30.228	.633

Cronbach's alpha measures consistency. The value can range between 0 and 1. There are different views on the level of the alpha value. Most frequently values of over .7 are acceptable (Homburg and Giering, 1998). Overall, Cronbach's alpha for the 34 items is in a satisfying range with .854. For six constructs the .7 level is met; for four constructs (loyalty, trust, value, sacrifice) the level is not exceeded. The disadvantage of Cronbach's alpha is that it is sensitive to the number of indicators and that it cannot be inference statistically assessed. Therefore, the next step is a confirmatory factor analysis to evaluate the measurement model.

Factor loadings for some items are less than .7 which is considered as lower limit by some authors. They suggest to drop the items lower than .7 from further analysis. This would involve reconsidering the constructs sacrifice, value, loyalty, trust, and enjoyment which also showed low Cronbach's alpha values. It strikes that some items show a loading higher than 1. If the items are correlated the coefficients have to be regarded as regression coefficients not as correlation coefficients as with orthogonal independent factors.

Significance of the construct is calculated in Mplus by providing the critical ratio. This is the unstandardized coefficient divided by the standard error. Indicator reliability shows the variance explained by an item. It should not be between 1.69 and -1.69 the values higher and lower are significant at p=.05. Some indicators show factor loadings below .7 but are still significant. This indicates that the variables are not appropriate to measure the latent construct.

R^2 is on average in a satisfying range but naturally lower for the items with lower factor loadings.

Now, Fornell and Larcker's (1981) approach is used to asses construct reliability, convergent validity, and discriminant validity. For the formulas please refer to chapter **Fehler! Verweisquelle konnte nicht gefunden werden.**. Construct reliability, also known as composite reliability uses the standardized loadings and measurement error for calculation. This will be carried out using an excel spreadsheet. Convergent validity is assessed with the average variance extracted (AVE). AVE is similar to construct reliability (CR), except that the standardized loadings are squared before being summed. AVE should be >.05 providing the variance due to measurement error is greater than the variance due to the construct.

Fornell and Larcker's (1981) approach to discriminant validity requires that AVE should exceed shared variance between the construct and all other variables in the model.

Table 27 summarizes the reliabilities of the final scales used in this survey. AVE is reported on the on-diagonal while squared correlations are reported on the off-diagonal. The AVE values in bold show squared correlations of the latent variables above the AVE value.

Table 27: Overview of the Reliability of the Scales

	CR	1	2	3	4	5	6	7	8	9	10
1 Qual	.94	.79									
2 Val	.76	.47	.51								
3 Sac	.79	.27	.55	.57							
4 Usab	.89	.46	.55	.40	.67						
5 Eff	.85	.46	.67	.48	.75	.66					
6 Enj	.88	.17	.03	.01	.13	.16	.71				
7 Tru	.80	.43	.34	.22	.53	.58	.14	.51			
8 Loy	.26	.43	.63	.37	.54	.75	.22	.52	.25		
9 Wom	.92	.45	.61	.27	.47	.67	.19	.48	.89	.79	
10 Pi	.86	.03	.03	.07	.03	.07	.12	.06	.02	.11	.67
CFI=.818; TLI=.942; RMSEA=.105; SRMR=.071; WRMR=1.721;

Average variance extracted should be higher than 50%. This criterion is not met by the construct loyalty and only just met by the constructs value, trust, and enjoyment.

The largest correlation (off-diagonal) is .89 and the lowest AVE is .25. Hence, the smallest on-diagonal is smaller than the largest off diagonal value. As a consequence discriminant and convergent validity are unsatisfactory. The measures should be reconsidered, in particular, the ones prooving problematic. These are the ones for loyalty, word-of-mouth, interpersonal influence, trust, and enjoyment.

There is room for improvement of the measurement model. First of all, it has to be reconsidered if the items with loadings <.7 should be retained for further analysis. They should be dropped and this might affect the too high RMSEA.

Secondly, the conceptualization of word-of-mouth and loyalty may need revision. Conceptualization of the loyalty construct included word-of-mouth as a component of loyalty combining the affective and cognitive components (Day, 1969). In service quality literature it is often one construct (Parasuraman, Zeithaml et al., 1994b; Parasuraman, Zeithaml et al., 2005). It was assumed that the word-of-mouth construct was more important in the setting of mobile services and treated as an individual construct different from loyalty.

Thirdly, the measurement of the trust concept seems problematic. Average variance extracted is at the lower end of acceptability. As more items that could be components of the trust concept were included in the questionnaire this dimension can be adapted. Also the ones of value and fun have to be reconsidered.

Before providing alternative models by modifying the measurement and structural model considering further theory from the field the original model is tested using dichotomous variables.

9.3.2 Using Dichotomous Variables for Model Test

Now, the same model is estimated using dichotomous variables. Since the scale used for measurement was a four point Likert scale ranging from *strongly agree* to *strongly disagree* the recoding to obtain dichotomous variables is straight forward. *Strongly agree* and *agree* are coded 0 instead of 1 and 2. *Rather disagree* and *disagree* are coded 1 instead of 3 and 4. The following graph illustrates the structural model and the measurement model. Each of the estimates is shown with the insignificant ones in brackets.

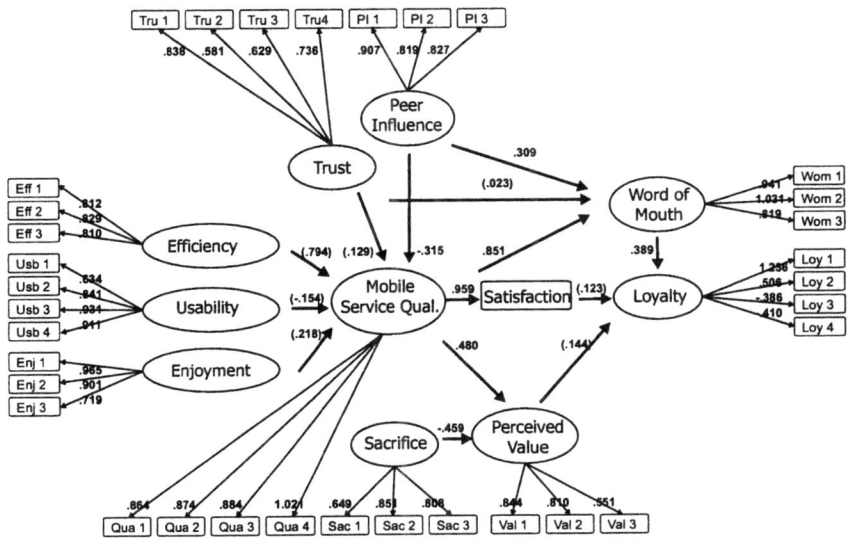

Figure 32: Full Research Model With Dichotomous Variables

Table 28 provides the fit indices for the model estimated with dichotomous variables. CFI should perform better for dichotomous outcomes than RMSEA. Generally use of SRMR is not recommended with binary outcomes and a cut off value of close to 1 is recommended for WRMR (Yu, 2002), which is obtained considering the value of 1.193. CFI and TLI are both above the required cut off value and RAMSEA is excellent with 0.53. Again, the model fit is good.

Table 28: Fit Indices for the Research Model with Dichotomous Measures

Fit Index	Obtained Level
CFI	.906
TLI	.950
RMSEA	.053
SRMR	.113
WRMR	1.193

Investigating the whole model it becomes evident that the previous one, with the four point Likert scale is not substantially superior. While the fit indices are improved for the model with binary data more path coefficients are insignificant. This time seven paths are insignificant. Again, those will not be included in the interpretation of the results and the model generally has to be interpreted carefully.

Table 29 provides an overview of the path coefficients, the critical ratio, and indicator whether the hypotheses could be supported or not.

Examining the antecedents of perceived service quality it is most noticeable that peer influence has a rather strong negative (-.315) effect while the other constructs are insignificant. One of the two antecedents with hypothesized effects on word-of-mouth was significant with a coefficient of .309, which is the one of peer influence. The influence of trust is insignificant and will therefore not be interpreted.

Again, the effect of perceived service quality is high on satisfaction (.959). Also the other two hypothesized paths of service quality on word-of-mouth (.851) and on perceived value (.480) are supported. The path between word-of-mouth and loyalty is again significant but with .389 a lot lower than with the four point scale data.

Finally, the following table summarizes all the hypotheses tested.

Table 29: Summary of the Findings of the Hypotheses Test

Hypotheses	Path Coefficient	Critical Ratio	Hypotheses Supported
H1 Eff -> Qua	.794	1.871	No
H2 Usa -> Qua	-.154	-.513	No
H3 Enj -> Qua	.218	1.761	No
H4 Val -> Loy	.144	1.344	No
H5 Sac -> Val	-.459	-3.901	Yes
H6 Pi -> Wom	.309	3.923	Yes
H7 Pi -> Qua	-.315	-1.983	Yes
H8 Tru -> Qua	.129	.537	No
H9 Tru -> Wom	.023	.166	No
H10 Wom -> Loy	.389	3.617	Yes
H11 Sat -> Loy	.123	1.906	No
H12 Qua -> Sat	.959	24.343	Yes
H13 Qua -> Wom	.851	7.238	Yes
H14 Qua -> Val	.480	5.511	Yes

The model fit is satisfactory. However, there are problems with insignificant path coefficients. As demonstrated in chapter 0 the measurement model will be evaluated for the dichotomized data in the following sub-chapter.

9.3.2.1 The Measurement Model Using Dichotomous Variables

The data set was recoded to assess the observables with dichotomous measures. The reason was to apply weaker assumptions regarding respondent's capability of discrimination.

For a better idea of the sample structure below figure provides the thresholds for the dichotomous variables. The figure shows that some items are close to the extreme points. Compared to Figure 31 depicting the thresholds for the categorical data with four intervals this one shows six more variables on the negative side of the scale.

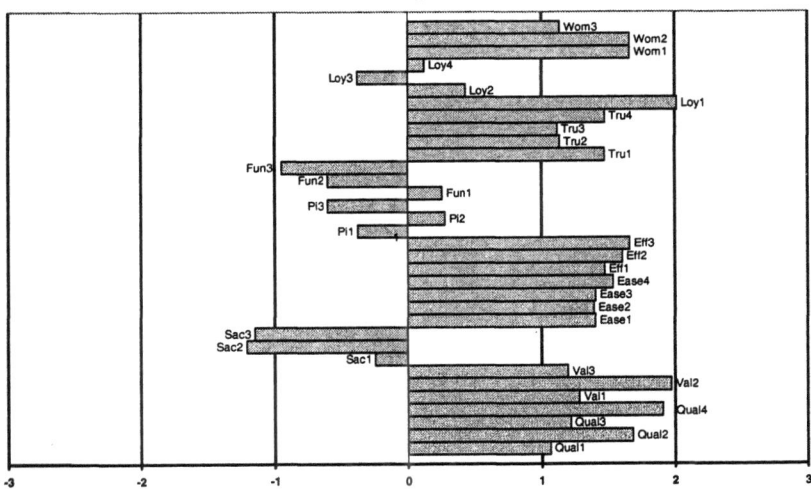

Figure 33: Thresholds of the Dichotomous Variables

The following table shows the Cronbach's alpha, factor loadings, critical ratio, and R^2 of each construct.

Table 30: Cronbach's Alpha, Factor Loading, Critical Ratio, and R^2

Item	Cronbach's Alpha	Factor Loading	Critical Ratio	R^2
Qual1	.751	.878	0.000	.771
Qual2		.874	16.606	.765
Qual3		.893	43.392	.797
Qual4		1.037	24.346	
Sat				
Val1	.538	.847	0.000	.718

Val2		.803	10.244	.645
Val3		.561	5.987	.315
Sac1	.494	.650	0.000	.422
Sac2		.852	7.867	.725
Sac3		.819	7.291	.671
Usab1	.701	.636	0.000	.405
Usab2		.833	7.865	.694
Usab3		.925	8.323	.856
Usab4		.920	6.829	.846
Eff1	.589	.806	0.000	.650
Eff2		.822	10.185	.676
Eff3		.801	8.651	.641
Pi1	.713	.914	0.000	.836
Pi2		.816	12.763	.665
Pi3		.829	11.845	.688
Enj1	.673	.945	0.000	.893
Enj2		.911	13.446	.830
Enj3		.729	9.925	.531
Tru1	.520	.840	0.000	.705
Tru2		.551	5.246	.304
Tru3		.638	6.413	.407
Tru4		.755	7.687	.570
Loy1	.246	1.221	0.000	
Loy2		.512	4.634	.262
Loy3		-.388	-3.350	.150
Loy4		-.418	-3.656	.174
Wom1	.735	.925	0.000	.856
Wom2		1.038	17.002	
Wom3		.0816	18.536	.665

With .721 the overall Cronbach's alpha for the items is a bit lower than for the original measurement scale. All in all, the Cronbach's alphas are lower for the dichotomous variables. Only four out of ten constructs exceed the .7 limit. Still, all the critical ratios indicate that the measures are significant. However, more factor loadings are above one, a phenomenon observed before. Word-of-mouth and loyalty, two constructs that showed problems earlier, are the ones with the noticeable values.

The following table shows construct reliability of the latent constructs. The diagonal provides the values for average variance extracted and the off diagonal shows the shared variance.

Table 31: Overview of the Reliability of the Scales

Construct	CR	1	2	3	4	5	6	7	8	9	10
1 Qual	.96	.85									
2 Val	.79	.43	.56								
3 Sac	.82	.22	.47	.61							
4 Usab	.90	.42	.26	.40	.70						
5 Eff	.85	.48	.62	.58	.65	.66					
6 Enj	.90	.17	.02	.00	.13	.18	.75				
7 Tru	.79	.43	.18	.22	.24	.48	.16	.50			
8 Loy	.31	.24	.22	.12	.18	.28	.11	.10	.52		
9 Wom	.95	.65	.61	.27	.29	.68	.24	.35	.41	.87	
10 Pi	.89	.00	.00	.00	.00	.06	.09	.01	.00	.08	.73
CFI=.900; TLI=.946; RMSEA=.053; SRMR=.110; WRMR=1.161;											

Fit indices improved by dichotomizing the variables. CFI, TLI and RMSEA now are within the required ranges.

Average variance extracted is higher for the dichotomous variables. The lowest one is the concept of trust just making it to .5. Discriminant validity appears to be better for the dichotomous variables, yet, some still show problems. The AVE values in bold show squared correlations of the latent variables above the AVE value.

9.3.3 Summary and Interpretation of Findings from the Model Test

The data fitted the hypothesized model well; however, due to insignificant path coefficients it only has limited explanatory power. There are various possible areas for improvement. Decisions for modifications have to be taken before continuing the analysis. The following issues have to be considered:

- Hardly any difference between the results for the model test with the data as collected and the binary data

Since there is no significant difference between the first (with data as collected) and the second (after re-coding with binary outcomes) analysis of the model it should be considered to continue with the analysis only using the binary data. As the difference is so little, this approach seems feasible and is followed.

- Definition of the construct loyalty
- Discriminant validity of loyalty and word-of-mouth

There are various definitions of the construct loyalty in literature. The approach taken, including behavioral and attitudinal components (Day, 1969) is further carried out. The attitudinal component, however, will now be word-of-mouth, an implicit form of attitudinal loyalty. The measurement of loyalty and word-of-mouth as separate constructs did not proove successful. This may explain the lack of discriminant validity. Past research shows conceptualizations of loyalty including word-of-mouth (Cronin, Brady et al., 2000; Parasuraman, Zeithaml et al., 2005); this view is now followed. The importance of the influence of interpersonal influence and a resulting degree of word-of-mouth in a mobile services setting was over estimated.

- Discriminant validity of efficiency and word-of-mouth

The revised conceptualization of the construct loyalty may lead to different path estimates and also to different results with regard to model evaluation. Therefore, no further changes with respect to the lack of discriminant validity between efficiency and word-of-mouth appear to be necessary.

- Definition of the construct trust

The average variance extracted of trust is at the lower end of acceptability in both model specifications. It is even lower than the problematic construct of loyalty. The operationalization included four items, two relating to trust in the company providing the service, two relating to trust into the functioning of the service. These two measures may be distinct from each other. Therefore, establishing two individual measures may be feasible. Literature shows that these two were measured individually previously. In services marketing Berry and Parasuraman (1991, 144) argue that *"customer company relationships require trust."* Trust into the company will now be an individual measure. The research from Internet shopping trust can be adapted for this survey. Various authors state that a lack of trust in the Web site (in this case trust in the functioning of a mobile service) could reduce the individual's desire to carry out transactions on the Web (Doney and Canon, 1997; Hoffman, Novak et al., 1999). Therefore, two dimensions of trust also established in literature are used.

- Low explanatory power of various constructs

Some path coefficients were not significant or extremely low in both model estimations. Consequently, their explanatory power has to be reconsidered. It may be advisable to drop some of those constructs for a more parsimonious model of the consumer behavior, which is of most interest for this research.

9.4 Alternative Models

Executing the improvements discussed in the previous chapter leads to two alternative models described in the following subchapters. The following analyses show this exploratory exercise.

9.4.1 Alternative 1

As discussed, there are problems with discriminant validity and construct specification. Therefore, these are reconsidered. Furthermore, some constructs are not further considered due to a lack of explanatory power. Since it is not explicitly the aim of this research to evaluate satisfaction it is also dropped from the model. The impact of value was considered to be more important for further research and therefore remains in the model. Figure 34 depicts the first alternative model.

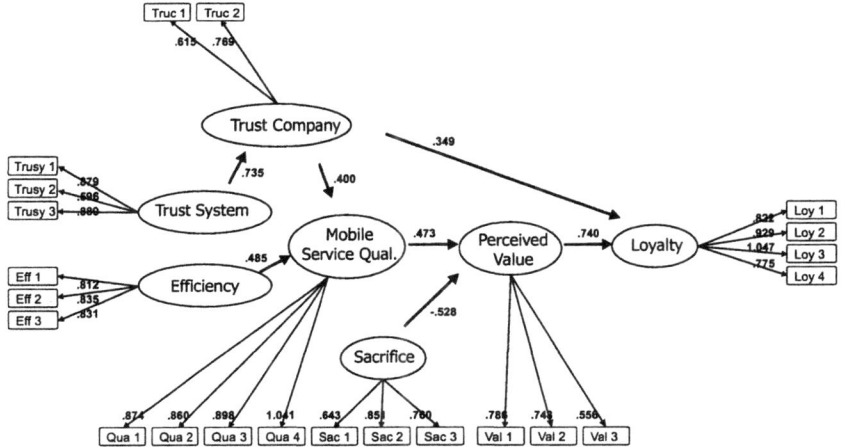

Figure 34: Alternative Model for Mobile Services Usage

The improvements lead to a better model fit, yet, there are issues involved with the new construct trust in the company and sacrifice. This is illustrated by below table. The paths are all significant and overall this is a satisfying model.

Table 32: Overview of the Reliability of the Scales for the Alternative Model

Construct	CR	1	2	3	4	5	6	7
1 Qual	.96	.85						
2 Val	.74	.55	.50					
3 Sac	.80	.26	.59	.57				
4 Eff	.87	.50	.59	.68	.68			
5 Trusys	.84	.43	.26	.14	.55	.64		
6 Trucom	.65	.44	.21	.08	.30	.54	.48	
7 Loy	.94	.61	.81	.44	.58	.40	.48	.81
CFI=.955; TLI=.980; RMSEA=.042; SRMR=.096; WRMR=.976;								

The evaluation of the measurement model lead to critical ratios generally higher than the average variance extracted. The average variance extracted should arrive at a value of over .50 which is achieved for all but one construct.

The following discussion on alternative two demonstrates another model solely focusing on the behavioral part of the initial research model.

9.4.2 Alternative 2

In this alternative the antecedents of service quality were excluded to focus on the behavioral part of the model, which is the main focus of this research project. The model was estimated once more. The fit measures are good and the measurement model is satisfying. Details are in Table 33.

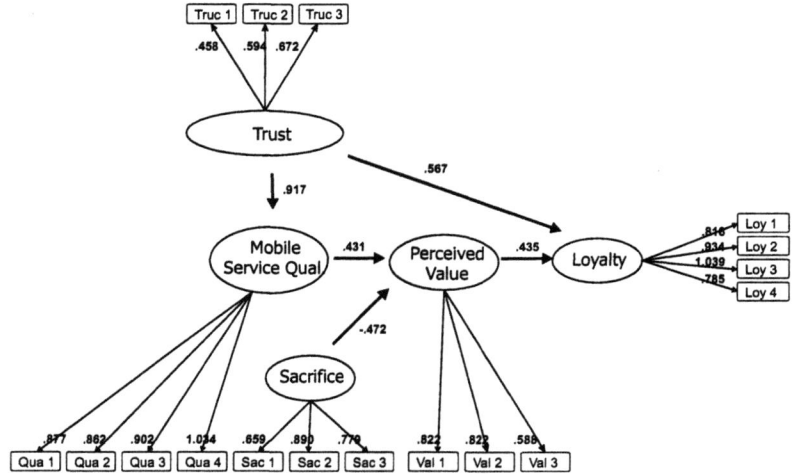

Figure 35: The Behavioral Model

The figure depicts the behavioral model. All of the path coefficients are significant. Overall the model did not change dramatically. Still, it is worth to mention that the effect of trust on the perceived quality appears to be very high with .917. Such high path coefficients indicate a lack of discriminant validity. This suggestion is not supported as the average variance extracted exceeds the shared variance of quality and trust. Further details can be found in the table below.

Table 33: Overview of the Reliability of the Scales for the Behavioral Model

Construct	CR	1	2	3	4	5
1 Qual	.96	.85				
2 Sac	.79	.22	.57			
3 Val	.82	.42	.45	.61		
4 Tru	.60	.84	.26	.40	.34	
5 Loy	.94	.64	.34	.63	.71	.81
CFI=.955; TLI=.980; RMSEA=.050; SRMR=.093; WRMR=1.036;						

Only one value is not excellent, the average variance extracted for trust. This is due to the fact that a very parsimonious model was sought for which allows further analyses such as latent class analysis and multiple group analysis.

The aim of these two methods is to find out if there are any group specific differences with regard to the mobile service usage behavior. The following chapter provides further details on the multiple group analyses and latent class analyses and, above that, presents the results.

9.5 Searching for Heterogeneity

The sample might vary greatly in mobile services usage, suggesting heterogeneity. This could be with regard to the mobile service usage pattern, lifestyle variables, or simply demographic background. Two approaches are used to detect these differences, multiple group analysis and latent class analysis. In the first one characteristics are identified a-priori that might cause sub-groups in the sample based on theory guided considerations. These are to be found in chapter 0. These moderators serve as grouping variables.

The second approach, latent class analysis, does not require a-priori assumptions. The analysis is carried out specifying how many classes should be built. Then, interpreting the output, it is decided which number of classes provides the best results. In the last step the different classes are described using further descriptive information from the data. The following two subchapters present multiple group analysis and latent class analysis.

9.5.1 Multiple Group Analysis

To detect differences between groups multi group analysis is carried out. The grouping variables are as hypothesized earlier, a person's:
- innovativeness
- age
- previous experience with a mobile service

There are three steps in multiple group analysis:
1. Fit of the model in all groups allowing all parameters to be free
2. Fit of the model in all groups holding factor loadings equal to test the invariance of the factor loadings
3. Fit of the model in all groups holding factor loadings and intercepts equal to test the invariance of the intercepts

MPlus can carry out multiple group analysis. The *grouping* option is used to identify the variable in the data set that contains the information on group membership. Two variations of the model command are used, *model* and *model + the grouping label*. Model describes the overall model estimated for each group. As default the factor loading measurement parameters are held equal across groups in order to specify measurement invariance. *Model* followed by a label describes the difference between the overall and the group specific model.

The command *meanstructure* in MPlus analyses means, thresholds and intercepts. The type *general* is the default but is changed to *meanstructure* for the third multi group analysis. When a model includes a mean structure, the intercepts of the continuous factor indicators are held equal across the groups as the default because they are measurement parameters. The intercepts of the factors are fixed to zero in group one and free to be estimated in the other groups as the default.

Sometimes estimation in Mplus does not run. Often an easy way out is to set the first item of a construct equal to one since the default setting in Mplus might occasionally not work. After doing so the estimations worked out fine for all the multiple group models.

Then the chi-Square difference is evaluated to see if there are significant group differences. However, Steenkamp and Baumgartner (1998) stress that one should go beyond the chi-square difference test as it suffers from the same problems as the chi-square test for evaluating model fit. Therefore, the change in fit indices such as RMSEA, TLI, and CFI are observed as they take into account model parsimony.

9.5.1.1 Innovativeness as Grouping Variable

As hypothesized, a person's innovativeness may lead to groups within a sample. The questionnaire included two statements (Goldsmith, 2001; Goldsmith and Hofacker, 1991) for domain specific innovativeness. A four point Likert scale was used for measurement. To yield groups big enough for the estimation the scale was not split in the middle but between *strongly agree* and *agree*.

As a result, the more innovative group one, counts 244 cases and group two, the not that innovative one, has 258 cases. To find out about the groups three models were estimated. The first one considering measurement non-invariance, the second one factor loading invariance and the third one factor loading and intercept invariance. Those are used to find out about configural invariance, metric invariance and factor variance/covariance.

Table 34: Chi-Square Change with Innovativeness as Grouping Variable

Model	Chi²	Chi² Diff	df	df Diff.	α	CFI	TLI	RMSEA
Measurement non-invariance	582		226			.854	.824	.079
Factor loading invariance	626	44	238	12	.001	.840	.817	.081
Factor loading and intercept invariance	649	23	250	12	.01	.836	.821	.080

The difference between the paths was evaluated using the unstandardized estimates from the second model test with invariant measurement model and paths freed to vary for the two groups. The chi-square change and the change of fit indices indicate model improvement or deterioration. However, these indicators have to be used cautiously. Model fit seems to first deteriorate and then to increase. The alpha values for chi-square change are highly significant.

The following table provides details on the path estimates for group one and two, the standard errors, the tolerance interval, and a statement whether there is a difference between the paths or not.

Table 35: Path Estimate Difference and Tolerance Interval with Innovativeness as Grouping Variable

Path	Group 2 Estimate	S.E.	Tolerance Interval	Group 1 Estimate	Difference
Tru -> Qual	2.742	0.814	1.114 - 4.370	2.663	No
Val -> Loy	0.109	0.030	0.049 - 0.169	0.251	Yes
Tru -> Loy	0.706	0.217	0.272 - 1.140	(0.207)	Yes
Sac -> Val	-0.718	0.209	(-1.136) - (-0.300)	-0.291	Yes
Qual -> Val	0.299	0.089	0.121 - 0.477	0.528	Yes

The table shows that there is a significant difference for four out of five paths. The estimates for group one do not lie in the tolerance interval calculated based on the group two estimates. The only path that shows no difference is the one between trust and quality.

Group one, the more innovative one, has specific characteristics. The effect of value on loyalty is higher. Also the effect of quality on value is higher. The sacrifice for using the service and its impact on value is lower for group one. The effect of trust on loyalty is lower for group one; however, this path is not significant. At the same time this path is high for group two. Obviously trust is the main determinant for loyalty for the less innovative group two. Value is less important in comparison. The effect of quality on value is strong for the innovators but less strong for the laggards.

In summary, trust is the main determinant for loyalty in group two. Also this group is less value driven with a very high influence of sacrifice on the value perception. M-parking causes more effort for the laggards. The quality value relationship is less important. Group one shows a stronger influence of quality on value and a weaker one for sacrifice. Exploring the determinants of loyalty for the innovators it is found that the effect of trust on loyalty is not significant but the effect of value is supported.

The following chapter shows the results for experience as a grouping variable.

9.5.1.2 Experience as Grouping Variable

It was hypothesized that users with more experience have different perceptions than those with less experience. Therefore, the respondents were asked about their usage habits and when they first used mobile services. Group two comprises users that use the service longer than a year while group one comprises those using mobile services less than a year.

The split of the sample is 289 respondents in group one and 213 respondents in group two. Again, to learn more about the group difference three models were estimated. The first one with measurement non invariance, the second one with factor loading invariance, and the third one with factor loading and intercept invariance. The following table illustrates the results of the model estimation.

Table 36: Chi-Square Change with Experience as Grouping Variable

Model	Chi²	Chi² Diff.	df	df Diff.	α	CFI	TLI	RMSEA
Measurement non-invariance	631		226			.830	.795	.085
Factor loading invariance	676	45	238	12	.001	.816	.790	.086
Factor loading and intercept invariance	682	6	250	12	.9	.819	.803	.083

The alpha coefficient for the difference between the estimation of the model with factor loading invariance and factor loading and intercept invariance is very high with .9. Again, there is no clear improvement of the fit indices, first all of they deteriorate and then they recover. Therefore, the tolerance interval is calculated.

The difference between the paths was evaluated using the estimates from the second model test with invariant measurement model and paths freed to vary for the two groups. Table 37 shows the path estimates for group one and two. All paths but one are significant.

Table 37: Path Estimate Difference and Tolerance Interval with Experience as Grouping Variable

Path	Group 2 Estimate	S.E.	Tolerance Interval	Group 1 Estimate	Difference
Tru -> Qual	2.812	0.886	1.040 - 4.584	3.382	No
Val -> Loy	0.072	0.036	0.000 - 0.144	0.121	No
Tru -> Loy	0.474	0.172	0.013 - 0.818	0.615	No
Sac -> Val	-0.612	0.18	(-0.972) - (-0.252)	-0.406	No
Qual -> Val	0.424	0.098	0.228 - 0.620	0.396	No

The analysis of the difference using the tolerance interval shows no difference of the path coefficient. Therefore, the next section continues with the third a-priori grouping variable which is age.

9.5.1.3 Age as Grouping Variable

Now the respondents' age is a grouping variable. Only respondents over 18 were considered in the analysis as one needs a driver's license to use the m-parking service. In case they were younger, they were excluded. The age of 30 is the cutting point with those under 30 in group one, which comprises 174 respondents. Group two, including the respondents who are 30 years old and above are 328.

Now the two groups are compared investigating the chi-square difference and the fit indices.

Table 38: Chi-Square Change with Age as Grouping Variable

Model	Chi^2	Chi^2 Diff.	df	df Diff.	α	CFI	TLI	RMSEA
Measurement non-invariance	643		226			.823	.787	.068
Factor loading invariance	670	27	238	12	.01	.817	.791	.085
Factor loading and intercept invariance	678	8	250	12	.8	.819	.803	.083

The alpha coefficient for the difference between the estimation of the model with factor loading invariance and factor loading and intercept invariance is .8. There is no clear improvement or deterioration of the fit indices. Thus, the tolerance interval is calculated.

The difference between the paths was evaluated using the estimates from the second model test with invariant measurement model and paths freed to vary for the two groups. Table 39 shows the path estimates for group one and two, the standard errors, the tolerance interval, and a statement whether there is a difference between groups or not.

Table 39: Path Estimate Difference and Tolerance Interval with Age as Grouping Variable

Path	Group 2 Estimate	S.E.	Tolerance Interval	Group 1 Estimate	Difference
Tru -> Qual	2.995	0.904	1.187 - 4.803	3.626	No
Val -> Loy	0.114	0.030	0.054 - 0.174	0.102	No
Tru -> Loy	0.583	0.190	0.203 - 0.963	0.679	No
Sac -> Val	-0.390	0.141	(-0.672)-(-0.108)	-0.681	Yes
Qual -> Val	0.437	0.081	0.275 - 0.599	0.334	No

Only one path differed for the age groups; the effect of sacrifice on value is bigger for the younger group. In summary, only one a-priori grouping variable detected group differences.

After investigating the groups after previously deciding upon grouping variables latent class analysis is carried out. This will show if the a-priori criteria are outstripped by other unknown variables.

9.5.2 Latent Class Analysis

This section presents the results of the latent class analysis, introduced by Lazarsfeld and Henry (1968), appliable with cross-sectional data, multiple items measuring one construct, and when the construct represents a latent class variable. These conditions all apply for this research. The aim of latent class analysis is the estimation of class probabilities and to relate class probabilities to covariates. This allows the classification of individuals into classes (posterior probabilities).

In Mplus the *classes* command is in the analysis part. The users stipulate the number of classes they want to find in order to compare the results of the various estimations. The model command is written down for as many classes as the user wants. In the model command one can specify if parts of the model (for instance the measurement model) should be held equal. It is also possible to let only parts of the structural model be freely estimated for all the classes wanted. For this project as much as possible is allowed to vary freely to obtain as much information on differences between users as possible.

Sometimes the procedure in Mplus does not work. It is helpful to introduce starting values and fix one item at one. Often only trying multiple starting values may lead to results. It is also advisable to select different sets of starting values and then compare the results. By making the comparison the results can be validated. This approach was taken.

Two of the paths had to be fixed, they were held equal across groups; otherwise the model would not converge. However, the measurement model was free to vary. Thus, the differences between the groups could be identified using all the information from the observables available. The paths between trust - loyalty and value – loyalty had to be held equal.

Since the measurement model was free to vary all the variables' information was included in the interpretation. The following graph depicts the means of the trust items for the four class solution. This is a good example for visualizing the group differences found.

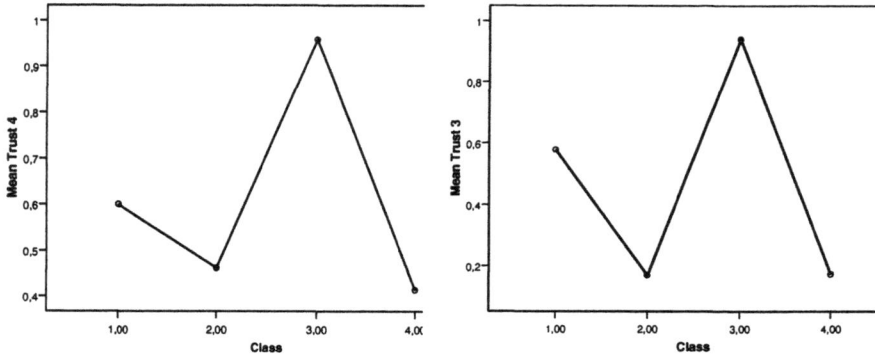

Figure 36: Means of the Four Classes for Two Trust Measures

The estimation of classes was carried out for a one class, two class, three class, four class, and five class solution. This allows a comparison of the loglikelihood change, the Akaike (AIC), Bayesian Information Criterion (BIC), and Entropy for the estimations. Table 40 provides an overview of these measures for the underlying model.

Table 40: Deciding on the Number of Classes

Number of Classes	Loglikelihood	AIC	BIC	Entropy
1 Class	-7292.772	14699.545	14940.005	NA
2 Classes	-7147.514	14451.028	14780.079	.769
3 Classes	-7080.042	14358.085	14775.726	.848
4 Classes	-6963.159	14166.318	14672.550	.718
5 Classes	-7197.024	14676.047	15270.870	.993

The BIC value is a good aid for choosing a model. Since the BIC value should be small, the four class solution is the preferred one. The quality of classification is assessed using the entropy value. An entropy value close to one indicates good classification in that many individuals have probabilities close to either one or zero with regard to class membership; thus, the five class solution is preferable. However, looking into the class counts and proportions shown in Table 41, the five class solution appears unbalanced. Further checking the class probabilities in Table 42 the values for the five class solutions are rather extreme. As learned from the literature, a scenario with great entropy but poorer fit according to BIC and the other way around is possible. Therefore, the measures provided above should be used cautiously.

The table below shows the counts for each of the classes and the respective proportions. Both the three class and the four class solution showed satisfactory and similar BIC and entropy values. Further examination of the results should be carried out. The figures in Table 41 indicate the three class solution is dominated by a rather big class two. When carrying out the analyses it was hoped that this rather dominant class would split up into two distinct classes. The four class solution shows a more balanced distribution. The big class two may have splited up when calculating the four class solution.

Table 41: Class Counts and Proportions

Number of Classes	Latent Classes	Counts	Proportions
1 Class	Class1	502	1
2 Classes	Class 1	77	.15339
	Class 2	425	.84661
3 Classes	Class 1	46	.09163
	Class 2	389	.77490
	Class 3	67	.13347
4 Classes	Class 1	95	.18924
	Class 2	65	.12948
	Class 3	168	.33466
	Class 4	174	.34661
5 Classes	Class 1	6	.01195
	Class 2	3	.00598
	Class 3	1	.00199
	Class 4	10	.01992
	Class 5	428	.96016

The following table depicts the latent class probability for most likely latent class membership by latent class. In the four class solution this would mean that the probability of membership in class one by class one is 87.7%. Generally the probability for most likely class membership is in a satisfactory range of well above 80% for each of the four classes. The five class solution arrives at a rather unbalanced result with one dominant and four rather small classes.

Table 42: Latent Class Probability for Most Likely Latent Class Membership (Row) by Latent Class (Column)

Number of Model	Latent Classes	1	2	3	4	5
2 Classes	Class 1	.883	.117			
	Class 2	.055	.945			
3 Classes	Class 1	.904	.069	.026		
	Class 2	.017	.955	.028		
	Class 3	.026	.090	.884		
4 Classes	Class 1	.877	.021	.090	.012	
	Class 2	.067	.816	.073	.043	
	Class 3	.102	.026	.836	.035	
	Class 4	.030	.050	.050	.870	
5 Classes	Class 1	.985	.000	.000	.000	.015
	Class 2	.000	.997	.000	.000	.003
	Class 3	.000	.000	1.000	.000	.000
	Class 4	.000	.000	.000	.889	.111
	Class 5	.000	.000	.000	.001	.998

Since the quality indicators for the four class solution is in a satisfactory range and superior to the results for the five class an three class solution this one will now be further analyzed.

When carrying out the analysis in MPlus one can impute a command which creates a data file with probabilities of class membership and a variable denoting the class membership. This data file is now combined with the original data file. This allows a more thorough look into the data, further interpretation of the results, and descriptions of the classes identified supported by the other variables available in the data set.

9.5.2.1 Model Differences

This section presents the main differences between the groups with regard to the alternative research model on the behavior of mobile services users. Since MPlus does not provide standardized estimates for this procedure the unstandardized ones are given below.

Table 43: Path Estimates and Standard Errors for the Four Classes

Path	Class	Estimate	S.E.	Estimate/S.E.
Tru -> Qual	Class 1	1.243	0.497	2.503
	Class 2	0.207	0.281	0.739
	Class 3	0.908	0.374	2.428
	Class 4	0.347	0.305	1.138
Val -> Loy	Class 1	0.416	0.139	2.996
	Class 2	0.416	0.139	2.996
	Class 3	0.416	0.139	2.996
	Class 4	0.416	0.139	2.996
Tru -> Loy	Class 1	0.232	0.188	1.235
	Class 2	0.232	0.188	1.235
	Class 3	0.232	0.188	1.235
	Class 4	0.232	0.188	1.235
Sac -> Val	Class 1	-0.631	0.254	-2.481
	Class 2	0.541	0.586	0.923
	Class 3	-1.116	0.673	-1.659
	Class 4	-0.140	0.293	-0.477
Qual -> Val	Class 1	0.930	0.181	5.136
	Class 2	0.088	0.310	0.283
	Class 3	0.338	0.260	1.301
	Class 4	0.044	0.090	0.485

When it comes to the path between trust and quality it is obvious that they are higher for classes one and three as opposed to the others. The same holds for the relationship between quality and value. The estimate for class one is particularly high with a rather small standard error which points toward a rather distinct difference. For this class the quality perception has a strong influence on the perceived value.

Analyzing the effect of sacrifice on value shows three negative paths for classes one, three and four. Originally it was hypothesized that the sacrifice one has to accept for consuming the service has a negative effect on the value perception. This is shown for three of the four classes in this book. For class two it is positive and relatively high with .541. However, looking into the standard errors, there is a rather high degree of variance and the results also indicate that they are only significant for the path between sacrifice and value of class two. The following section characterizes the four classes.

9.5.2.2 Distinct Characteristics of the Classes Identified

After adding the class labels to the data set, analysis and further description of the classes is possible.

First of all, it was interesting to explore whether the variables assumed to have a grouping effect in the multiple group analysis also proved to show differences in the latent classes. Therefore, cross tabulations were calculated using the grouping variables from the previous chapter. The analysis shows that for gender the results of the chi-square test were not significant and, therefore, the relationship between the latent classes and gender could not be established.

Next, the relationship between innovativeness and the latent classes was tested. The results were significant at (p <.000). Class three seems to be the most innovative one with over 66% stating they tend to be rather innovative while classes two and four are least innovative with a proportion of 55% and 60% stating they tend to be less innovative.

The investigation of a difference with regard to the experience of using a mobile service assumed that early and late adopters differ in their perceptions and behavior. Therefore, the cross tabulation is carried out with the duration of mobile service usage and the latent classes. The results are not significant (p<.422).

Finally, the results suggest that there is no relationship between age and the latent class membership.

To further describe the class differences and learn more about the variables' difference across the classes an ANOVA was calculated. The results are highly significant at the p<.000 level and p<.002 for one variable. Now based on the findings the classes are described in further detail.

9.5.2.2.1 Class One: Mobile Service Skeptics

The first class can be named *Mobile Service Skeptics*. They have the worst quality perception and the value is not perceived to be high either. Generally there was not much variance between the sacrifice for using the service across classes, they all considered it as high. However, for this class the sacrifice seems to be highest. This class does not trust the service that much. Altogether, it ranks third when it comes to trust. Also with regard to loyalty this class seems rather reserved. In summary, they can be characterized as reluctant to actions in favor of the service.

The *Mobile Service Skeptics* is the most reluctant class of mobile service users. They use it least frequently. Also, with regard to innovativeness this class ranks in the middle. These individuals appear to be technology laggards. They use few new mobile services and are not particularly innovative. Therefore, the may have a poor perception of the services in general, do not trust the service and are not loyal.

Their average age is 33.5. 57% earn below 2000 Euros and the majority of 38% have A-levels and 38% even a higher university degree. With 22% self-employed they are strongest in this class. The distribution of earnings differs little from all the other classes.

9.5.2.2.2 Class Two: Undecided Users

This class lies between the extremes of class one and four in the perceived quality of the service. In this class the path for sacrifice and value was positive, however, not significant. Nevertheless the sacrifice was higher in this class than in classes three and four (see below *the Cautious Innovators and the Mobile Service Lovers*). Class two, the *Undecided Users*, is ranking second with regard to trust. The users of this class consider the service as trustworthy. As opposed to classes one and three this one tends to show more loyalty.

The *Undecided Users* use the service second most frequently after class four *(The Mobile Service Lovers)*. This class appears to be stuck in the middle. It is more innovative and tends to show a higher degree of loyalty and trust. However, they are not totally convinced of the service's quality and value.

57% of this class are male and 42% are female. Even more (37%) completed a university degree and 35% completed their A-levels. There are fewer students in this class and most (64%) are employees. This class tends to earn less than others with 68% under 2000 Euro.

9.5.2.2.3 Class Three: Cautious Innovators

Generally class three, the *Cautious Innovators*, perceives the quality and value of the mobile service lower than the other classes. For this class the sacrifice for using the service is lower than for others. These respondents have the least trust in the service; they are cautious mobile services users. The *Cautious Innovators* are also the least loyal, which might relate to the lack of trust since it is hypothesized that trust has a positive effect on loyalty.

Class three ranks third in mobile service usage. A significant relationship between service usage frequency and class membership could be established. This class is the most innovative and likes to experiment with new technologies. They like to try a service because it is new and they enjoy trying innovations but they are not loyal users.

Among the *Cautious Innovators* are more males (68%) than females (32%). They hold the biggest share of people who obtained A-levels (43%) and a university degree (35%). This is in line with the argument that innovators are better educated than others. They also hold a big share of self-employed with 15%. With 24% earning above 2000 Euro and 13% above 3000 Euro they also hold the biggest share of high earners.

9.5.2.2.4 Class Four: Mobile Service Lovers

The *Mobile Service Lovers* have the best perception of the quality of the service. They rank it as extremely high. Also the perceived value in relation to effort, time and cost is high for this class. The sacrifice for using the service is least for this class. The *Mobile Service Lovers* are the most trusting respondents. They consider the service and the service provider as trustworthy. Also, the loyalty of this class appears to be very high. This may be supported by the high trust proneness.

The members of this class are the heaviest users of the mobile service. Most use it more than once a week. This class experiments the least with new technologies. This may reflect the high degree of loyalty. This class, however, has least tendency to try innovations. Once they found a service that works for them, they stick with it and are loyal users, not seeking the most innovative alternative.

In this class 63% are male and 37% are female. The distribution of education is even in this class. There are a bit fewer who hold a university degree among the m-service lovers. However, the share of self-employed is big (17%) in this class. As in all other classes the employees are the biggest class.

10 CONCLUSIONS

This book introduces the concept of m-commerce, the m-commerce value chain, mobile services, and their characteristics after an exploratory research phase. Expert interviews yielded insights into m-commerce. Among those interviewees were executives and researchers from nine countries. The findings from literature review were combined with those of the expert interviews for a more detailed view of the unexplored field of mobile services.

The research model was developed based on a thorough analysis of literature. Theoretic building blocks for the model development were service quality and diffusion of innovations theory.

An online survey was carried out to collect data for the model test. 502 m-parking users completed the questionnaire and were included in the analysis.

The estimation of the model used the software package Mplus. The tool was convenient and flexible for SEM, multiple group analysis, and latent class analysis.

The survey lead to a number of interesting results. The hypothesized causal relationships were supported to a great extent. Before focusing on the implications for industry and researchers the main results are summarized and discussed. Among those methodological findings are the quality of performance measures, binominal data as measures for SEM, theory guided formulation of alternative models, different segments within the sample, and the quality of a-priory criteria for segmentation as opposed to latent class analysis.

- *Performance only measures are well suited as measurement instrument*

The literature review for the measurement instrument pointed toward performance and expectations oriented measures. However, empirical evidence showed that performance measures had equal or more explanatory power and, moreover, a lower degree of complexity for the interviewees. This suggests that performance evaluation is preferred over the expectations measure (Cronin and Taylor, 1992; Cronin and Taylor, 1994). The analyses of the data showed that performance measures perform well when measuring service quality.

- *Binary data are not inferior to other scales in the model test*

The analysis of the measurement model showed that the four point Likert scale is not significantly better than the one with binominal variables. Therefore, it was decided to continue the analysis with binominal variables.

- *The alternative models*

Guided by theory, alternative models were defined. The explanatory power of those and the quality of the measurement model can be considered as satisfying.

The results indicate that trust and value have the most influence on the users' loyalty. Value is primarily determined by the quality of a service and the sacrifice. Trust in the service also has an effect on the perceived quality of the service. The modification of the measurement model and the structural model proved successful since often more parsimonious models lead to good results. Furthermore, this more parsimonious model allowed for using advanced statistical techniques.

The data fit the model well, however, the perceptions of customers, may differ in the sample, therefore, those differences were explored.

- *Customer groups differ in their perceptions of the mobile services*

The results show that there are differences among user groups with regard to their perceptions of mobile services. A significant relationship between group membership and duration of service use leads to the suggestion that concepts like loyalty and value change over time. A limitation is that the constructs are measured at one point in time. A longitudinal analysis would give a dynamic perspective. There is literature suggesting a dynamic perspective of loyalty (see chapter 0 for Oliver's dynamic loyalty perspective). Trying to put such dynamic constructs in a static statistical tool has problems. Such tools, broadly used in research, only allow for snapshots in time, and as a result, important relationships may be overlooked or remain undetected. In a model allowing for different phases (introduction, maturity, decline) the strength of the paths and even the differences in the measurement model could be assessed. These claims, however, are in contrast to the methods available for researchers. To account for dynamic effects a longitudinal study, potentially over a couple of years, would be necessary. In addition, an extension of the model would increase complexity and further hamper data collection and model estimation. Researchers have to make trade-offs between the explanatory power of their models and the complexity of reality.

- *A-priori criteria in multiple group analysis only partly lead to satisfying results as opposed to latent class analysis*

Three a-priori criteria for detecting groups in the sample, chosen based on theory and expert views, were innovativeness, experience, and age. All of those were found to be significant moderators in other surveys. The results of the multiple group analyses show that only innovativeness is a significant grouping variable. The others did not arrive with significant group differences.

Latent class analyses supported these findings. Out of the three hypothesized moderators only innovativeness proved to be significant in the chi-square test. However, the power of latent class analysis is to find specific groups and flag the cases according to their group membership. As a result, it is possible to carry out further analyses with all the variables. The discovery of groups with such a high degree of variation and internal consistency would not have been possible by solely relying on a-priory criteria. However, it is the researchers' task to find descriptive variables that best characterize the classes found through the analysis. This may be challenging if not enough data is collected or if, besides from the measurement and structural model, the other variables do not differ between the classes.

The classes identified and the fact that the differences between the classes are essential leads to implications for industry. These are discussed later in this chapter. First, implications for future research are discussed.

10.1 Implications for Future Research

There is potential for future research in a number of areas. First, the merit of replication of this study is discussed. Second, potential research avenues lie in model extensions and refinement. Further methodology driven potentials in the area of linearity and non-linearity of relationships between constructs are highlighted. Finally, this section presents trends in marketing research and advanced methods to gain customer data.

- *Replication of the study in different industries and regions*

A replication of this study including different services and validating the measurement instrument would be worthwhile. The path estimates may differ according to the type of service. It even was different for various groups of customers as the latent class analysis showed. Different services could include fun-oriented ones – such as ring tone and music download – to efficiency-driven ones like mobile ticketing.

Given a strong theoretic development, future research in this field should develop and test hypotheses related to mobile services. In addition to conceptual work that investigates evolving mobile technologies, the role of culture merits investigation. There is a growing field of research drawing upon culture as an influencing factor in the adoption of technology (Bayarmaa and Boalch, 1997; Ford, Connelly et al., 2003; Loch, Straub et al., 2003; Norris, 2001). It would be interesting to compare the results of the survey across countries to find if the diffusion rates are driven and influenced by cultural backgrounds or the development of the mobile communication industry in a specific country. In Japan mobile services are very common and provided via the i-mode platform and in Scandinavia mobile Java games are popular. Both services, yet, are not that well accepted in the German speaking countries.

Furthermore, as innovativeness was a useful grouping variable, cultural background must be reconsidered. In Japan, a country with a collectivistic background, individual services such as wallpapers to personalize the mobile phone are common. This may lead to the suggestion that Hofstede's (1980) dimensions are not fully applicable with technological innovations. Furthermore, when replicated, the wording and applicability of the measurement instrument must be considered and adapted.

- *Model extensions and refinements*

Despite thorough literature review, the operationalization of constructs was not without problems. An attempt to reduce skewness of the data by reconsidering the anchors of the scales is valuable. It would be possible to just measure the perceptions of dissatisfaction with anchors of *very bad* to *bad* instead of the anchors *very good* to *very bad*. They usually lead to a majority of the answers in the areas *very good* and *good*. Another approach would be to dichotomize the data and split right after the *very good* statement to have two groups nearly equal in size instead of the same answer-content (positive vs. negative answer), as it was done in this survey.

Furthermore, there are potential constructs worth investigating and including in a future survey, such as variety seeking and pricing.

Finally, the models presented only show a potential relationship between the constructs. Often literature contradicted with regard to the direction of paths. Therefore, the researcher had to decide which to choose. Thus, alternative nested and non-nested models could be formulated and tested. Those may lead to equally good or even better results.

- *Linearity and non-linearity of relationships between constructs*

A third area of interest for future research lies in methodology development. Research on satisfaction has shown that the relationship between attribute level performance and overall satisfaction is non-linear. Matzler et al. (2004) argued for the existence of different types of asymmetric relationships (basic factors, excitement factors) as well as linear factors (performance factors). Mittal et al. (1998) argue for a stronger impact of negative attribute performance on overall satisfaction and the results also indicate that a re-examination of the consequent behaviors of satisfaction is necessary. Klein and Muthén (2004) developed a statistically efficient and practicable estimation method for structural equation models with multiple latent-non linear effects that outperform other available approaches with respect to efficiency. Multiple interaction and quadratic effects can be estimated with Klein's Quasi-ML method.

- *Future trends in consumer research – or – learning to better listen to the customer*

For innovative services, such as mobile services, the input of customers for new service development is essential. The following paragraphs present innovative methods to explicitly and implicitly collect data about the customer. The mobile medium and mobile services in general lend themselves to new methods for consumer research.

Dahan and Hauser (2002) investigate the capabilities of communication and information technologies for rapid and inexpensive consumer input at the stages of the new product development process. The authors explore six Web-based methods of consumer input. The suggested Information Pump allows consumers to interact with each other in an early stage of product development, the Fast Polyhedral Adaptive Conjoint Estimation allows screening larger numbers of product features, in User Design consumers can design their own products, Virtual Concept Testing provides a platform where product development teams can actually build a product. Finally, a stock-market simulation called Securities Trading of Concepts identifies winning concepts. Compared to traditional market research techniques, new media tools are more cost-efficient and accurate, and allow listening to the customer and product developers at different stages of the innovation process (Dahan and Hauser, 2002).

A similar approach provides toolkits for customers to co-create their preferred product. By shifting innovation tasks from the manufacturer to the consumer (von Hippel, 1986), new product development can benefit from creative users and meet increasingly heterogeneous needs more effectively (von Hippel, 1999).

These two approaches are effective, but depend on the customers' willingness to provide information. At the same time, customers voluntarily post valuable comments and product information on the Internet, but companies lack the ability to detect, analyze and use this source of information. Brohman et al. (2003) introduce the concept of net-based customer service systems (NCSS), which operate either directly via a Web browser, PDA or cell phone, or indirectly via a service representative or agent. Data-mining tools support NCSS in analyzing data for market segmentation, profiling customers and matching services. Most companies only store data on past transactions. Few companies have achieved a broader view and lack the capability to collect data across multiple brands and products. Third-party travel firms such as Travelocity and online mass merchandisers like Amazon are very effective at collecting this type of data. Amazon.com's recommender system collects data related to multiple brands by using adaptive filters to identify up-selling and cross-selling opportunities (Zhang and Im, 2002).

A contribution by Watson (2004) highlights customer-managed interaction (CMI), a new service model on the horizon. This would solve the argument about who owns customer data; in this approach the customers themselves would have total control over information about their purchases and preferences. When applying CMI the customer has full control of the content, mode and timing of data exchange for all vendor and service encounters (Watson, 2004).

These methods of consumer input are innovative, but costly and time consuming. Above that the decision makers are dependent on consumers willingly providing information.

Shardanand and Maes (1995) introduce a social filtering system that automates *"word-of-mouth"* recommendations for music. The technology helps navigate the abundance of digital information. Social (collaborative) filtering systems recommend items for the user's consumption based on similarities of the values assigned by other people and the customer's own values. Thus, individual recommendations are generated through correlations in the value structure. The Internet is not just a medium to obtain information, it also facilitates communication between customers who want to share their experiences with a particular good or service. Web sites provide virtual opinion platforms where consumers can tap articulations on products, services and companies (Henning-Thurau and Walsh, 2003). As traditional methods of gathering and aggregating this type of information are time consuming and expensive, automated tools may be a welcome alternative.

In the field of mobile services an abundance of Internet platforms exist where customers exchange comments and views. These may be used in the future for implicit collection of customers' perceptions on services.

10.2 Implications for Industry

Mobile applications are in their early stages and should develop rapidly in the next years. Two complementary factors help fuel this evolution. On the one hand, user acceptance of the services continues to evolve as the devices become simpler and easier to use (Pedersen and Herbjorn, 2003). On the other hand, manufacturers continue to improve and enhance their mobile devices. Usage constraints such as a small display and inconvenient handling should diminish as the technology continues to improve. For special user groups or companies with unique requirements, specially adapted mobile devices are also possible. This accounts for the case explored in this survey. Police men, for instance, are equipped with special PDAs that enable them to check if drivers paid the fee and in case they did not executive authority can directly print a parking ticket.

There are lessons learned for industry that derive from this book. Among those are insights in the importance of the quality - value - loyalty chain, the importance of different segments and targeting them accordingly, and offering mobile services in general. Those main findings with implications for industry are now explained in more detail.

- *Importance of the quality - value - loyalty chain*

The nature of the mobile medium lets users send and receive messages at any time. This is one value proposition of mobile services. It is important that services deliver value since this has a positive impact on the users' satisfaction and usage behavior (Zeithaml, 1988). The results show that the well established causal chain between perceived quality – value – loyalty is supported in this research. The second effect on loyalty is caused by trust. For companies this implies that it is important to offer reliable services to create some degree of loyalty. The value proposition is also important. Customers may be unaware of the value resulting from using a service.

For companies it is important to know which factors it can influence to create a high degree of loyalty among their customers and, hence, reduce churn. Among the factors they can influence is the reliability of the system they provide the service with. This is SMS in the case of m-parking. Since SMS is a best effort service, in other words it is not certain the message arrives if there is to much traffic on the network, providers have to offer some service recovery if messages delivery fails. Furthermore, since fines are heavy if the driver does not pay the parking fee on time, trust in proper service delivery is necessary. Payment is settled via the network operator so, once more, trust into security of the payment process are important.

- *Importance of different segments and tailored targeting of those*

As learned from the above discussion there are segments that vary with respect to the perceptions and are distinct from each other. Accordingly, it is important to target them differently. Knowledge from companies' customer relationship tools can do this. Often companies do not fully use the possibilities for customer relationship management or are not aware of their potential.

The *Mobile Service Skeptics* have a bad quality and value perception and perceive the sacrifice as highest. Companies have to either highlight the advantages of using mobile services for this group or accept that this group is not using many of the services offered. Generally, this class is not innovative and does not use many other mobile services either.

The *Undecided Users* could be turned into more heavy users by companies. The members of this class are loyal users who trust the service but their perceptions of quality and value of the service are not excellent. This class hast to be convinced of the quality and value of the service. This may even result in more heavy use or even in use of new services. Hence, additional revenue can be generated for companies.

The *Cautious Innovators* trust the service the least. They have to be convinced that the service works well. In some cases they even may have made bad experience when using the service. Companies should have effective service recovery procedures not to loose an unsatisfied customer. At the same time it is this class that counts the most innovators. Those are important when new services are introduced. They should be in favor of the service provider as they may serve as opinion leaders for less innovative users.

The last class, the *Mobile Service Lovers*, should be retained by the providers. There might be potential that those buy even more services in the future. They have a high degree of trust in the service providers; therefore, they are loyal and may consume future offers as well.

- *Offer of mobile services – competition within the market*

Most mobile network operators offer mobile services. For some they are important as their corporate strategy is to be the market leader in innovations and customer orientation. Their aim is to increase average revenue per customer through cross selling mobile services. Some offer packages with flat fees that allow usage of certain services at no additional charge. Others use pay-as-you-go pricing models for services such as m-parking. Most of the services are only provided by one network operator. Mobilkom for instance, offers the Vodafone life! platform as gateway to most mobile services, while t-mobile offers services on t-zones. Such offers, however, are always limited to a specific customer group, the company's own customers. Generally, it is important to mention that services that should have sustainable success should be offered independent of the platform or net-provider. This is the case with m-parking. Initially not all providers collaborated but by now it is possible to pay for parking fees via SMS, no matter what network operator one has.

Yet, the customer service offered and the specific pricing model of voice communication is the main basis of decision, which provider to pick. In a couple of years, however, the offer of mobile services may be a further factor considered when deciding on which network provider to pick.

11 REFERENCES

12Snap (2001). Mobile-Marketing Aktion für Wella Design erfolgreich: 55.000 Mobile Küsse, Press Release (7 December 2001).

724.Solutions (2001). Commerce Goes Mobile. San Francisco, 724 Solutions Inc.

AFX.Asia (2001). NTT DoCoMo Seeks Approval for Push-Type Advertising, Info service on iMode. 2001.

Agarwal, R. and Karahanna, E. (2000) "Time Flies When You're Having Fun: Cognitive Absorption and Beliefs About Information Technology Usage", MIS Quarterly, 24, (4), 665-694.

Agarwal, R. and Prasad, A. (1998) "A Conceptual and Operational Definition of Personal Innovativeness in the Domain of Information Technology", Information Systems Research, 9, (2), 204-215.

Aijzen, I. (1991) "The Theory of Planned Behavior", Organizational Behavior and Human Decision Processes, 50, 179-211.

Ajzen, I. (2001) "Nature and Operation of Attitudes", Annual Review of Psychology, 52, 27-58.

Ajzen, I. and Fishbein, M. (1980), Understanding Attitudes and Predicting Social Behavior, Prentice-Hall, Englewood Cliffs, NJ.

Alderson, W. (1965), Dynamic Marketing Behavior, Richard D. Irwin, Inc., Homewood, Illinois.

AMA (1985) "AMA Board Approves New Marketing Definition", Marketing News, 19, (1), 1.

Anckar, B. and D'Incau, D. (2002). Value-Added Services in Mobile Commerce: An Analytical Framework and Empirical Findings from a National Consumer Survey. 35th Hawaii International Conference on System Sciences, Hawaii, USA, IEEE.

Anderson, J. C. and Gerbing, D. W. (1984) "The Effect of Sampling Error on Convergence, Improper Solutions, and Goodness-of-Fit Indices for Maximum Likelihood Confirmatory Factor Analysis", Psychometrika, 49, (1984), 155-173.

Anderson, J. C. and Gerbing, D. W. (1988) "Structural Equation Modeling in Practice: A Review and Recommended Two-Step Approach", Psychological Bulletin, 103, (3), 411-423.

Anderson, J. C. and Narus, J. A. (1990) "A Model of Distributor Firm and Manufacturer Firm Working Partnerships", Journal of Marketing, January, 42-58.

Arbuckle, J. L. and Wothke, W. (1999), Amos 4.0 User`s Guide, SmallWaters Corporation, Chicago.

Arndt, J. (1967) "Role of Product-Related Conversations in the Diffusion of a New Product", Journal of Marketing Research, 4, (August), 291-295.

Au, A. K.-M. and Enderwick, P. (2000) "A Cognitive Model on Attitude Towards Technology Adoption", Journal of Managerial Psychology, 15, (4), 266-282.

Babbie, E. (1998), The Practice of Social Research, Wadsworth Publishing Company, Belmont.

Backhaus, K., Erichson, B., Plinke, W. and Weiber, R. (2000), Multivariate Analysemethoden: Eine anwendungsorientierte Einführung, Springer, Berlin.

Bagozzi, R. (1981a) "Evaluating Structural Equation Models with Unobservable Variables and Measurement Error: A Comment", Journal of Marketing Research, 18, (3 (August)), 357-381.

Bagozzi, R. (1982) "Evaluating Structural Equation Models With Unobservable Variables and Measurement Error", Journal of Marketing Research, 18, 375-381.

Bagozzi, R. and Yi, Y. (1988) "On the Evaluation of Structural Equation Models", Journal of the Academy of Marketing Science, 16, (1), 74-94.

Bagozzi, R. P. (1980), Causal Models in Marketing, Wiley, New York.

Bagozzi, R. P. (1981b) "Attitudes, Intentions, and Behavior: A Test of Some Key Hypotheses", Journal of Personality and Social Psychology, 41, (4), 607-627.

Bagozzi, R. P. (1994). Measurement in Marketing Research: Basic Principles of Questionnaire Design. Principles of Marketing Research. R. P. Bagozzi. Cambridge, Blackwell Publishers.

Bahattacherjee, A. (2000) "Acceptance of e-Commerce Services: The Case of Electronic Brokerages", IEEE Transactions of Systems, Man, and Cybernetics, 30, (4), 411-420.

Balasubramanian, S., Peterson, R. A. and Jarvenpaa, S. L. (2002) "Exploring the Implications of M-Commerce for Markets and Marketing", Journal of the Academy of Marketing Science, 30, (4), 348-361.

Bandura, A. (1986), Social Foundations of Thought and Action, Prentice-Hall, Englewood Cliffs, NJ.

Bansal, H. S. and Voyer, P. A. (2000) "Word-of-Mouth Processes within a Services Purchase Decision Context", Journal of Service Research, 3, (2), 166-177.

Barnes, S. and Vidgen, R. (2000). WebQual: An Exploration of Web-site Quality. Eigth European Conference on Information Systems, Vienna.

Barnes, S. I. and Vidgen, R. (2001a). An Evaluation of Cyber-Bookshops: The WebQual Method., International Journal of Electronic Commerce, M.E. Sharpe Inc. 6: 11.

Barnes, S. J. (2002a) "The Mobile Commerce Value Chain: Analysis and Future Developments", International Journal of Information Management, 22, 91-108.

Barnes, S. J. (2002b) "Wireless Digital Advertising: Nature and Implications", International Journal of Advertising, 21, 399-420.

Barnes, S. J. and Vidgen, R. (2003) "Measuring Web Site Quality Improvements: A Case Study of the Forum on Strategic Management Knowledge Exchange", Industrial Management and Data Systems, 103, (5/6), 297.

Barnes, S. J. and Vidgen, R. T. (2001b) "Assessing the Effect of a Web Site Redesign Initiative: An SME Case Study", International Journal of Management Literature, 1, (113-126).

Barnes, S. J. and Vidgen, R. T. (2001c). Assessing the Quality of Auction Web Sites. Hawaii International Conference on Systems Sciences, Maui, Hawaii.

Barnes, S. J. and Vidgen, R. T. (2001d) "An Evaluation of Cyber-Bookshops: The WebQual Method", International Journal of Electronic Commerce, 6, 6-25.

Barnes, S. J. and Vidgen, R. T. (2002) "An Integrative Approach to the Assessment of E-Commerce Quality", Journal of Electronic Commerce Research, 3, (3).

Barwise, P., Elberse, A. and Hammond, K. (2002). Marketing and the Internet: A Research Review, London Business School. 2003.

Barwise, P. and Strong, C. (2002) "Permission-based Mobile Advertising", Journal of Interactive Marketing, 16, (1), 14-24.

Bass, F. M. (1969) "A New Product Growth Model For Consumer Durables", Management Science, 15, (5), 215-228.

Battacherjee, A. (2000) "Acceptance of E-Commerce Services: The Case of Electronic Brokerages." IEEE Transactions on Systems, Man and Cybernetics, 30, 411-420.

Baumgartner, H. and Homburg, C. (1996) "Applications of Structural Equation Modeling in Marketing and Consumer Research: A Review", International Journal of Research in Marketing, 13, 139-161.

Bayarmaa, B. and Boalch, G. (1997). A Preliminary Model of Internet Diffusion within Developing Countries. Third Australian World Wide Web Conference (AusWeb-97), Lismore, Australia, Southern Cross University.

Bentler, P. M. (1990) "Comparative Fit Indexes in Structural Models", Psychological Bulletin, 107, (2), 238-246.

Bentler, P. M. and Bonett, D. G. (1980) "Significance Tests and Goodness of Fit in the Analysis of Covariance Structures", Psychological Bulletin, 88, (588-606).

Bentler, P. M. and Chou, C.-P. (1987) "Practical Issues in Structural Modeling", Sociological Methods and Research, 16, (1), 87-117.

Berekoven, L., Eckert, W. and Ellenrieder, P. (2001), Marktforschung: Methodische Grundlagen und Praktische Anwendung, Gabler, Wiesbaden.

Berry, L. L. (1993) "Playing Fair in Retailing", Arthur Anderson Retailing Issues Newsletter, March, (5), 2.

Berry, L. L. and Parasuraman, A. (1991), Marketing Services, Free Press, New York.

Berry, L. L. and Parasuraman, A. (1997) "Listening to the Customer - The Concept of a Service-Quality Information System", Sloan Management Review, Spring, 65-76.

Berry, L. L., Parasuraman, A. and Zeithaml, V. A. (1988) "The Service-Quality Puzzle", Business Horizons, (September-October), 35-43.

Bhattacherjee, A. (2002) "Individual Trust in Online Firms: Scale Development and Initial Test", Journal of Management Information Systems, 19, (1), 211-241.

Bitner, M. J. (1990) "Evaluating Service Encounters: The Effects of Physical Surroundings and Employee Responses", Journal of Marketing, 54, (April), 69-82.

Bollen, K. (1989), Structural Equations with Latent Variables, John Wiley & Sons, New York.

Bollen, K. and Long, S. (1993), Testing Structural Equation Models, Sage Publications, Newbury Park.

Bolton, R. N. and Drew, J., H. (1991) "A Multi-Stage Model of Customers' Assessments of Service Quality and Value", Journal of Consumer Research, 17, (4), 375-378.

Bone, P. F. (1992) "Determinants of Word-of-Mouth Communications During Product Consumption", Advances in Consumer Research, 19.

Bone, P. F. (1995) "Word-of-Mouth Effects on Short-term and Long-term Product Judgements", Journal of Business Research, 32, 213-223.

Boomsma, A. (1982). The Robustness of LISREL Against Small Sample Sizes in Factor Analysis Models. Systems Under Indirect Observation: Causality, Structure, Prediction. K. G. J. H. Wold. Amsterdam, North-Holland: 149-173.

Boomsma, A. (1985) "Nonconvergence, Improper Solutions, and Starting Values in LISREL Maximum Likelihood Estimation", Psychometrika, 52, 345-370.

Borden, N. H. (1942), Economic Effects of Advertising, Irvin, Chicago: Illinois.

Bossert, J. L. (1991), Quality Function Deployment: A Practitioner's Approach, ASQC Quality Press; M. Dekker, New York.

Brady, M. K., Cronin, J. J. and Brand, R. R. (2002) "Performance-only Measurement of Service Quality: A Replication and Extension", Journal of Business Research, 55, 17-31.

Brohman, M. K., Piccoli, G., Watson, R. T. and Parasuraman, A. (2005). NCSS Process Completeness: Construct Development and Preliminary Validation. 38th Hawaii International Conference on System Sciences, Big Island, Hawaii, IEEE Computer Society.

Brohman, M. K., Watson, R. T., Piccoli, G. and Parasuraman, A. (2003) "Data Completeness: A Key to Effective Net-Based Customer Service Systems", Communications of the ACM, 46, (6), 47-51.

Browne, M. W. (1984) "Asymptotically Distribution-Free Methods for the Analysis of Covariance Structures", British Journal of Mathematical and Statistical Psychology, 37, 62-83.

Browne, M. W. and Cudeck, R. (1993). Alternative Ways of Assessing Model Fit. Testing Structural Equation Models. K. Bollen and S. Long. Newbury Park, CA, Sage Publications.

Bruner, G. C. and Kumar, A. (2005) "Explaining Consumer Acceptance of Handheld Internet Devices", Journal of Business Research, 58, 553-558.

Bughin, J., Lind, F., Stenius, P. and Wilshire, M. (2001) "Mobile Portals: Mobilize for Scale", The McKinsey Quarterly, 38, (2), 118-127.

Burt, R. S. (1973) "Confirmatory Factor-Analytic Structures and the Theory Construction Methods", Sociological Methods and Research, 2, 131-187.

Burt, R. S. (1976) "Interpretational Confounding of Unobserved Variables in Structural Equation Models", Sociological Methods and Research, 5, (3-52).

Burzynski, M. H. and Bayer, D. J. (1977) "The Effect of Positive and Negative Prior Information on Motion Picture Appreciation", Journal of Social Psychology, 101, 215-218.

Byrne, B. M. (2001), Structural Equation Modelling With AMOS: Basic Concepts, Applications and Programming, Erlbaum, Mahwah.

Campbell, D. T. and Fiske, D. W. (1959) "Convergent and Discriminant Validation by the Multitrait-Multimethod Matrix", Psychological Bulletin, 56, 81-105.

Chau, P. Y. K. and Hu, P. J. H. (2002) "Investigating Healthcare Professional's Decisions to Accept Telemedicine Technology: An Empirical Test of Competing Theories", Information & Management, 39, (4), 297-311.

Chen, Q., Clifford, S. J. and Wells, W. D. (2002) "Attitude Toward the Site II: New Information", Journal of Advertising Research, March/April, 33-45.

Chen, Q. and Wells, W. D. (1999) "Attitude Toward the Site", Journal of Advertising Research, September/October, 27-37.

Chen, S. Y. and Macredie, R. D. (2005) "The Assessment of Usability of Electronic Shopping: A Heuristic Evaluation", International Journal of Information Management, 25, (6), 516-532.

Cheung, W., Chang, M. K. and Lai, V. S. (2000) "Prediction of Internet and World Wide Web Usage at Work: A Test of an Extended Triandis Model", Decision Support Systems, 30, (1), 83-100.

Childers, T., Carr, C., Peck, J. and Carson, S. (2001) "Hedonic and Utilitarian Motivations for Online Retail Shopping Behavior", Journal of Retailing, 77, (Winter 2001), 511-535.

Chircu, A. M. and Kauffman, R. J. (2001). Digital Intermediation in Electronic Commerce - the eBay Model. E-Commerce and V-Business. S. J. Barnes and B. Hunt. Oxford, Butterworth-Heinemann.

Choi, S.-Y., Stahl, D. and Whinston, A. (1997), The Economics of Electronic Commerce, Macmillan, New York.

Chordiant (2003). Chordiant 5 Marketing Director Suite.

Chou, C.-P. and Bentler, P. M. (1995). Estimates and Tests in Structural Equation Modeling. Structural Equation Modeling. Concepts, Issues, and Applications. R. H. Hoyle. Thousand Oaks, CA, Sage: 37-55.

Clifton. R. (2002) "Brands and Our Times", Journal of Brand Management, 9, (3), 157-161.

CNN (2003). New Switching Rules May Trash Millions of Cell Phones, Cable Network News.

Cohn, M. (2001) "Short, Sweet Talk", Internet World.

Compeau, D. R. and Higgins, C. A. (1995) "Computer Self-Efficacy: Development of a Measure and Initial Test", MIS Quarterly, June, 189-211.

Compeau, D. R., Higgins, C. A. and Huff, S. (1999) "Social Cognitive Theory and Individual Reactions to Computing Technology: A Longitudinal Study", MIS Quarterly, 23, (2), 145-158.

Copeland, M. T. (1923) "Relation of Consumer's Buying Habits to Marketing Methods", Harvard Business Review, 1, 282-289.

Corigliano, M. A. and Baggio, R. (2004). Mobile Technologies Diffusion in Tourism: Modelling a Critical Mass of Adopters in Italy. 11th International Conference on Information Technology and Travel and Tourism, Cairo, Egypt.

Cronin, J. J., Brady, M. K., Brand, R. R., Hightower, R. and Shemwell, D. J. (1997) "A Cross-sectional Test of the Effect and Conceptualization of Service Value", Journal of Services Marketing, 11, (6), 375-391.

Crorin, J. J., Brady, M. K. and Hult, G. T. M. (2000) "Assessing the Effects of Quality, Value, and Customer Satisfaction on Consumer Behavioral Intentions in Service Environments", Journal of Retailing, 76, (2), 193-218.

Cronin, J. J. and Taylor, S. A. (1992) "Measuring Service Quality: A Reeximation and Extension", Journal of Marketing, 56, (July), 55-68.

Cronin, J. J. and Taylor, S. A. (1994) "SERVPERF Versus SERVQUAL: Reconciling Performance-Based and Perceptions-Minus-Expectations Measurement of Service Quality", Journal of Marketing, 58, (January), 125-131.

Cudeck, R. and Browne, M. W. (1983) "Cross-validation of Covariance Structures", Multivariate Behavioral Research, 18, 147-167.

Dabholkar, P. A. and Bagozzi, R. P. (2002) "An Attitudinal Model of Technology-Based Self-Service: Moderating Effects of Consumer Traits and Situational Factors", Journal of the Academy of Marketing Science, 30, (3), 184-201.

Dahan, E. and Hauser, J. R. (2002) "The Virtual Customer", Journal of Product Innovation Management, 19, 332-353.

Damanpour, F. (1991) "Organizational Innovation: A Meta-analysis of Effects of Determinants and Moderators", Academy of Management Journal, 34, (3), 555-590.

Datamonitor (2000). US Mobile Market Worth USD 1.2 Billion by 2005, Datamonitor. 2000.

Davis, F. D. (1989) "Perceived Usefulness, Perceived Ease of Use, and User Acceptance of Information Technology", MIS Quarterly, 13, (3), 319-340.

Davis, F. D., Bagozzi, R. and Warshaw, P. (1989) "User Acceptance of Computer Technology: A Comparison of Two Theoretical Models", Management Science, 35, (8), 982-1003.

Davis, F. D., Bagozzi, R. P. and Warshaw, P. R. (1992) "Extrinsic and Intrinsic Motivation to Use Computers in the Workplace", Journal of Applied Social Psychology, 22, (14), 1111-1132.

Day, G. S. (1969) "A Two Dimensional Concept of Brand Loyalty", Journal of Advertising Research, 9, (September), 29-35.

DeLone, W. H. and McLean, E. R. (1992) "Information Systems Success: The Quest for the Dependent Variable", Information Systems Research, 3, (1), 60-95.

Deree&Company (2004). Green Star Field Doc, Deree & Company. 2004.

Devine, A. and Holmqvist, S. (2001). Mobile Internet Content Providers and their Business Models. Stockholm, Master Thesise, at the Royal Institute of Technology.

DeZoysa, S. (2002) "Japan and Europe - Worlds Apart?" Telecommunications, International Edition, 36, (3), 38-40.

Dick, A. S. and Basu, K. (1994) "Customer Loyalty: Toward an Integrated Conceptual Framework", Journal of the Academy of Marketing Science, 22, (Winter), 99-113.

Dickinger, A., Haghirian, P., Murphy, J. and Scharl, A. (2003). An Investigation and Conceptual Model of SMS Marketing. 37th Annual Hawaii International Conference on System Sciences (HICSS-37), Hawaii Big Island, IEEE.

Dickinger, A., Heinzmann, P. and Murphy, J. (2004). Mobile Environmental Applications. 38th Hawaii International Conference on System Sciences (Hicss), Hawaii, Big Island, IEEE.

Dickinger, A., Murphy, J. and Scharl, A. (2003). Web Coverage of Mobile Marketing by the Fortune Global 500. 4th International Working With e-business Conference, Perth, Australia.

Dodds, W. B. (1991) "In Search of Value: How Price and Store Name Influence Buyers' Product Perceptions", Journal of Services Marketing, 5, (Summer), 27-36.

Dodds, W. B., Monroe, K. B. and Grewal, D. (1991) "The Effects of Price, Brand, and Store Information on Buyers' Product Evaluations", Journal of Marketing Research, 28, (August), 307-319.

Donabedian, A. (1987) "Commentary on Some Studies of the Quality of Care", Health Care Financing Review, Annual Supplement.

Donabecian, A. (1991). Reflections on the Effectiveness of Quality Assurance. Striving for Quality in Health Care - an Inquiry into Policy and Practice. R. H. Palmer, A. Donabedian and G. J. Povar. Ann Arbor, Michigan, Health Administration Press.

Donabedian, A. (2003), An Introduction to Quality Assurance in Health Care, Oxford University Press, Oxford.

Doney, P. M. and Canon, J. P. (1997) "An Examination of the Nature of Trust in Buyer-Seller Relationships", Journal of Marketing, 61, (2), 35-51.

Döring, N. (2002) "1 Bread, Sausage, 5 Bags of Apples I.L.Y. - Communicative Functions of Text Messages (SMS)", Zeitschrift für Medienpsychologie, 14, (3), 118-128.

Durix, J.-F. (2003) "Revolutionising Mobile Payments", Card Technology Today, 15, (10), 10-11.

Durlacher.Research (2000). Internet portals. London, Durlacher Research.

Dwyer, R., Schurr, P. and Oh, S. (1987) "Developing Buyer Seller Relationships", Journal of Marketing, 51, (April), 11-27.

Engel, J. E., Blackwell, R. D. and Kegerreis, R. J. (1969) "How Information is Used to Adopt an Innovation", Journal of Advertising Research, 9, 3-8.

Ericsson (2000). Wireless Advertising. Stockholm, Ericsson Ltd.

Erlancson, C. and Ocklind, P. (1998) "WAP - the Wireless Application Protocol", Ericsson Review, 75, (4), 150-153.

Fazio, R. H. and Zanna, M. (1978) "Attitudinal Qualities Felating to the Strength of the Attitude Behavior Relationship", Journal of Experimental Social Psychology, 14, (4), 398-408.

Feldmann, S. P. and Spencer, M. C. (1965). The Effect of Personl Influence in the Selection of Consumer Services. AMA American Marketing Association, Chicago, American Marketing Association.

Fishbein, M. and Ajzen, I. (1975), Belief, Attitude, Intention and Behavior: An Introduction to Theory and Research, Addison-Wesley, Reading, MA.

Flynn, L. R. and Goldsmith, R. E. (1993) "A Validation of the Goldsmith and Hofacker Innovativeness Scale", Educational and Psychological Measurement, 53, 1105-1116.

Forc, D. P., Connelly, C. E. and Meister, D. B. (2003) "Information Systems Research and Hofstede's Culture's Consequences: An Uneasy and Incomplete Partnership", IEEE Transactions on Engineering Management, 50, (1), 8-25.

Fornell, C. (1983) "Issues in the Application of Covariance Structure Analysis: A Comment", Journal of Consumer Research, 9, (4), 443-448.

Fornell, C. and Larcker, D. F. (1981) "Evaluating Structural Equation Models with Unobservable Variables and Measurement Error", Journal of Marketing Research, 18, (1), 39-50.

Funk, J. L. (2002). The Product Life Cycle Theory and Product Line Management: The Case of Mobile Phones. Kobe, Kobe University, Research Institute for Economics and Business Administration.

Funk, J. L. (2003). Key Technological Trajectories and the Expansion of Mobile Internet Applications. Stockholm Mobility Roundtable, Stockholm.

Gefen, D. (2002) "Reflections on the Dimensions of Trust and Trustworthiness among Online Consumers", The Data Base for Advances in Information Systems, 33, (3), 38-53.

Gefen, D. and Straub, D. (1997) "Gender Differences in the Perception and Use of E-Mail: An Extension to the Technology Acceptance Model", MIS Quarterly, 21, (4), 389-400.

Gefen, D. and Straub, D. (2000) "The Relative Importance of Perceived Ease of Use in IS Adoption: A Study of E-Commerce Adoption", Journal of the Association for Information Systems, 1.

Gerbing, D. W. and Anderson, J. C. (1988) "An Updated Paradigm for Scale Development Incorporating Unidimensionality and Its Assessment", Journal of Marketing Research, 25, (2), 186-192.

Geser, H. (2002). Towards a Sociological Theory of the Mobile Phone, http://socio.ch/mobile/t_geser1.htm.

Gibson, S. (1997) "Spin a Website to Attract More Customers", PC Week, March 31, 1997.

Gilbert, A. L. and Kendall, J. D. (2003). A Marketing Model for Mobile Wireless Services. 37th Hawaii International Conference on System Sciences (HICSS-37), Hawaii, USA, IEEE.

Godin, S. (2001), Unleashing the Idea Virus, Hyperion, New York.

Goldsmith, R. E. (2001) "Using the Domain Specific Innovativeness Scale to Identify Innovative Internet Consumers." Internet Research: Electronic Networking Applications and Policy, 11, 149-158.

Goldsmith, R. E. and Hofacker, C. (1991) "Measuring Consumer Innovativeness", Journal of the Academy of Marketing Science, 19, (209-221).

Golem.de (2002). Mobiles Marketing ist noch ein Wunschtraum. 17 August 2001.

Goodhue, D. L. (1995) "Understanding User Evaluations of Information Systems", Management Science, 41, (12), 1827-1844.

Goodhue, D. L. and Thompson, R. L. (1995) "Task-Technology Fit and Individual Performance", MIS Quarterly, June, 213-236.

Green, P. E., Tull, D. S. and Albaum, G. (1988), Research for Marketing Decisions, Prentice Hall, Englewood Cliffs, New Jersey.

Grewal, D., Monroe, K. B. and Krishnan, R. (1998) "The Effects of Price-Comparison Advertising on Buyers' Perceptions of Acquisition Value, Transaction Value, and Behavioral Intentions", Journal of Marketing, 62, (April), 46-59.

Grönroos, C. (1978) "A Service-Oriented Approach to Marketing of Services", European Journal of Marketing, 12, (8), 588-601.

Grönroos, C. (1982). Strategic Management and Marketing in the Service Sector. Helsinki, Swedish School of Economics and Business Administration, Report No. 8.

Grönroos, C. (1984) "A Service Quality Model and its Marketing Implications", European Journal of Marketing, 18, (4), 36-44.

Grönroos, C. (1988) "Service Quality: The Six Criteria of Good Perceived Service Quality", Review of Business, 9, (3), 10-13.

Grönroos, C. (2001) "The Perceived Service Quality Concept - a Mistake?" Managing Service Quality, 11, (3), 150-152.

Grossnickle, J. and Raskin, O. (2001), Handbook of Online Marketing Research, McGraw-Hill, New York.

Gutman, J. (1982) "A Means-End Chain Model Based on Consumer Categorization Processes", Journal of Marketing, 46, (Spring), 60-72.

Haddon, L. (1997). Communications on the Move: The Experience of Mobile Telephony in the 1990s., European Commission, Sweden, Telia AB.

Haghirian, P., Dickinger, A. and Kohlbacher, F. (2004). Adopting Innovative Technology - A Qualitative Study among Japanese Mobile Consumers. 5th International Working with e-Business Conference (WeB-2004), Perth, Australia, Cowan University.

Hair, J. F., Anderson, R. E., Tatham, R. L. and Black, W. C. (1995), Multivariate Data Analysis with Readings, Prentice-Hall, Englewood Cliffs, NJ.

Hair, J. F., Busch, R. P. and Ortinau, D. J. (2000), Marketing Research: A Practical Approach for the New Millennium, McGraw-Hill Higher Education, United States.

Haley, R. I. and Baldinger, A. L. (1991) "The ARF Copy Research Validity Project", Journal of Advertising Research, 31, (2), 11-32.

Hamblen, M. (2000). Hello, This is a Wireless Ad. 2000.

Hanson, W. (2000), Principles of Internet Marketing, South-Western College Publishing, Cincinnati.

Harrison, D. A., Mykytyn, J. and Riemenschneider, C. K. (1997) "Executive Decisions About Adoption of Information Technology in Small Business: Theory and Empirical Tests", Information Systems Research, 8, (2), 171-195.

Harrison-Walker, J. L. (2001) "The Measurement of Word-of-Mouth Communication and an Investigation of Service Quality and Customer Commitment as Potential Antecedents", 4, 1, (60-75).

Hart, P. D. (2000). The Wireless Marketplace in 2000. Washington DC, Peter D. Hart Research Associates.

Hartmann, J. and Büppelmann, R. (2001). Siemens End-User Survey Europe. Munich, Siemens AG.

Hashimoto, Y. (2002) "The Spread of Cellular Phones and Their Influence on Young People in Japan", Review of Media, Information and Society, The Institute of Socio-Information and Communication Studies, The University of Tokyo, 7, 97-110.

Hashimoto, Y., Komatsu, A., Kurihara, M., Hanme, K. and Kashyap, A. (2001) "Communication Behavior of the Youth in the Metropolitan Area: Focusing on the Usage of the Internet, Mobile Phone and Short Message System", The Research Bulletin of the Institute of Socio-Information and Communication Studies, The University of Tokyo, 2001, (16), 94-210.

Hassanein, K. and Head, M. (2003). Ubiquitous Usability: Exploring Mobile Interfaces within the Context of a Theoretical Model. 15th Conference on Advanced Information Systems Engineering (CAiSE 2003), Klagenfurt, Austria.

Henning-Thurau, T. and Walsh, G. (2003) "Electronic Word-of-Mouth: Motives for and Consequences of Reading Customer Articulations on the Internet", International Journal of Electronic Commerce, 8, 51-74.

Herr, P. M., Kardes, F. R. and Kim, J. (1991) "Effects of Word-of-Mouth and Product Attribute Information on Persuasion: An Accessibility-Diagnosticity Perspective", Journal of Consumer Research, 17, (4), 454-456.

Hinde, S. (2003) "Spam: The Evolution of a Nuisance", Computer & Security, 22, (6), 474-478.

Hoffman, D. L. and Novak, T. P. (1996) "Marketing in Hypermedia Computer-mediated Environments: Conceptual Foundations", Journal of Marketing, 60, (July), 50-68.

Hoffman, D. L., Novak, T. P. and Peralta, M. (1999) "Building Consumer Trust Online", Communications of the ACM, 42, (4), 80-85.

Hofstede, G. (1980), Culture's Consequences, international differences in work-related values, Sage Publications, Newbury Park.

Holbrook, M. (1996) "Customer Value - A Framework For Analysis and Research", Advances in Consumer Research, 23, 138-142.

Holbrook, M. (1999), Consumer Value - A Framework for Analysis and Research, Routledge, London.

Homburg, C. and Baumgartner (1998). Beurteilung von Kausalmodellen: Bestandsaufnahme und Anwendungsempfehlungen. Die Kausalanalyse: Instrument der empirischen betriebswirtschaftlichen Forschung. Stuttgart, Schäffer-Poeschel.

Homburg, C. and Giering, A. (1998). Konzeptionalisierung und Operationalisierung komplexer Konstrukte: Ein Leitfaden für die Marketingforschung. Die Kausalanalyse: Instrument der empirischen betriebswirtschaftlichen Forschung. L. Hildebrandt and C. Homburg. Stuttgart, Schäffer-Poeschel.

Homburg, C. and Giering, A. (2001) "Personal Characteristics as Moderators of the Relationship Between Customer Satisfaction and Loyalty - An Empirical Analysis", Psychology and Marketing, 18, (1), 43-66.

Homburg, C. and Hildebrandt, L. (1998). Die Kausalanalyse: Bestandsaufnahme, Entwicklungsrichtungen, Problemfelder. Die Kausalanalyse: Instrument der empirischen betriebswirtschaftlichen Forschung. L. Hildebrandt and C. Homburg. Stuttgart, Schäffer-Poeschel.

Hu, L.-T. and Bentler, P. M. (1995). Evaluating Model Fit. Structural Equation Modeling: Concepts, Issues and Applications. R. H. Hoyle. Thousand Oaks, Sage Publications: 76-99.

Hu, P. J., Chau, P. Y. K., Lui Seng, O. R. and Yan Tam, K. (1999) "Examining the Technology Acceptance Model Using Physicians Acceptance of Telemedicine Technology", Journal of Management Information Systems, 16, 91-112.

Hulland, J., Chow, Y. H. and Lam, S. (1996) "Use of Causal Models in Marketing Research: A Review", International Journal of Research Marketing, 13, 181-197.

Hung, S.-Y., Ku, C.-Y. and Chang, C.-M. (2003) "Critical Factors of WAP Services Adoption: An Empirical Study", Electronic Commerce Research and Applications, 2, (1), 42-60.

Hyvönen, K. and Repo, P. (2004). Diffusion of Mobile Services in Finland. 3rd International Business Information Management (IBIMA) Conference, Cozumel, Mexico.

Igbaria, M., Iivari, J. and Mafagahh, H. (1995) "Why do Individuals Use Computer Technology? A Finnish Case Study", Information & Management, 29, (5), 227-238.

Jacoby, J. and Chestnut, R. W. (1978), Brand Loyalty Measurement and Management, Wiley, New York.

Jacoby, J. and Kyner, D. (1973) "Brand Loyalty vs. Repeat Purchasing Behavior", Journal of Marketing Research, X, (1-9).

Jarvenpaa, S. L. (1989) "The Effect of Task Demands and Graphical Format on Information Processing Strategies", Management Science, 35, (3), 285-303.

Jarvenpaa, S. L. and Todd, P. A. (1997) "Consumer Reactions to Electronic Shopping on the World Wide Web", International Journal of Electronic Commerce, 1, (2), 59-88.

Jayanti, R. K. and Fosh, A. K. (1996) "Service Value Determination: An Integrative Perspective", Journal of Hospitality and Leisure Marketing, 34, (4), 5-25.

Jee, J. and Lee, W.-N. (2002) "Antecedents and Consequences of Perceived Interactivity: An Exploratory Study", Journal of Interactive Advertising, 3, (1).

Johnston, C. (2000). Presentation to the Emerging Payments Technology Group. 2003.

Jones, N. (2001). Multimedia Mobile Messaging: A Marketer's Delight, Gartner Group Research Note (October 23).

Jones, T. and Sasser, E. W. (1995) "Why Satisfied Customers Defect", Harvard Business Review, 73, (November/December), 88-99.

Jöreskog, K. G. and Sörbom, D. (1981). Analysis of Linear Structural Relationships by Maximum Likelihood and Least Squares Methods, Research Report. Sweden, University of Uppsala.

Jöreskog, K. G. and Wold, H. (1982). The ML and PLS Techniques for Modeling with Latent Variables: Historical and Competitive Aspects. Systems Under Indirect Observation, Part 1. K. G. Jöreskog and H. Wold. Amsterdam, North-Holland: 263-270.

Kalluviayil, S. (2001). Owning the Wireless Customer Experience.

Kannan, P., Chang, A. and Whinston, A. (2001). Wireless Commerce: Marketing Issues and Possibilities. 34th Hawaii International Conference on System Sciences (HICSS-34), Hawaii, USA, IEEE.

Karahanna, E., Straub, D. and Chervany, N. (1999) "Information Technology Adoption Across Time: A Cross-Sectional Comparison of Pre-Adoption and Post-Adoption Beliefs", MIS Quarterly, 23, (2), 183-213.

Katz, E. and Lazarsfeld, P. (1955), Personal Influence, The Free Press, Glencoe, IL.

Katz-Stone, A. (2001). Wireless Revenue: Ads Can Work. 2001.

Kelloway, E. K. (1998), Using LISREL for Structural Equation Modelling, Sage Publications, London.

Kepper, G. (1996), Qualitative Marktforschung, Deutscher Universitätsverlag, Wiesbaden.

Kettinger, W. J. and Lee, C. C. (1994) "Perceived Service Quality and User Satisfaction with the Information Services Function", Decision Sciences, 25, (5), 737-766.

Kim, S. and Stoel, L. (2004) "Apparel Retailers: Website Quality Dimensions and Satisfaction", Journal of Retailing and Consumer Services, 11, 109-117.

Kleijnen, M., de Ruyter, K. and Wetzels, M. (2004) "Consumer Adoption of Wireless Services: Discovering the Rules, while Playing the Game", Journal of Interactive Marketing, 18, (2), 51-61.

Kleijnen, M., Ruyter, K. d. and Wetzels, M. (2002). Customer Adoption of Wireless Entertainment Services. 31st Annual EMAC Conference, Braga, Portugal.

Kleijnen, M., Wetzels, M. and de Ruyter, K. (2004) "Consumer Acceptance of Wireless Finance", Journal of Financial Services Marketing, 8, (3), 206-217.

Klein, A. G. and Muthén, B. O. (2004) "Quasi Maximum Likelihood Estimation of Structural Equation Models with Multiple Interaction and Quadratic Effects", Journal of the American Statistical Association, Under Review.

Kotler, P., Jain, D. C. and Maesincee, S. (2002), Marketing Moves, Harvard Business School Press, Boston.

Koufaris, M. (2002) "Applying the Technology Acceptance Model and Flow Theory to Online Consumer Behavior", Information Systems Research, 13, (2), 205-223.

Kumar, V., Aaker, D. A. and Day, G. S. (2002), Essentials of Marketing Research, John Wiley & Sons Inc., Hoboken.

Kwon, H. S. and Chidambaram, L. (2000). A Test of the Technology Acceptance Model: The Case of Cellular Phone Adoption. 34th Annual Hawaii International Conference on System Sciences, Hawaii, IEEE.

LaRose, R., Mastro, D. and Easiton, M. S. (2001) "Understanding Internet Usage: A Social-cognitive Approach to Uses and Gratifications", Social Science Computer Review, 19, 395-413.

Lazarsfeld, P. and Henry, N. W. (1968), Latent Structure Analysis, Houghton Mifflin, New York.

Lederer, A. L., Maupin, D. J., Sena, M. P. and Zhuang, Y. (2000) "The Technology Acceptance Model and the World Wide Web", Decision Support Systems, 29, 269-282.

Legris, P., Ingham, J. and Collerette, P. (2003) "Why do People Use Information Technology? A Critical Review of the Technology Acceptance Model", Information & Management, 40, 191-204.

Lehmann, H. and Lehner, F. (2001). The Next E Will be M - but M-What? Making Sense of Mobile Applications - a Critical Note. 15th Bled Electronic Commerce Conference, Bled, Slovenia.

Leung, L. and Wei, R. (1999) "Seeking News via the Pager: An Expectancy-value Study", Journal of Broadcasting & Electronic Media, 43, 299-315.

Liao, S., Shao, Y. P. and Wang, H. (1999) "The Adoption of Virtual Banking: An Empirical Study", International Journal of Information Management, 19, (1), 63-74.

Lin, C. A. (1996) "Looking Back: The Contribution of Blumler and Katz's Uses of Mass Communication to Communication Research", Journal of Broadcasting & Electronic Media, 40, 574-582.

Ling, R. (2001). It is in. It Doesn't Matter if You Need It or Not, Just That You Have it. Fashion and the Domestication of the Mobile Telephone among Teens in Norway., Working Paper, Telenor R&D. Oslo, Norway.

Loch, K. D., Straub, D. and Kamel, S. (2003) "Diffusing the Internet in the Arab World: The Role of Social Norms and Technological Culturation", IEEE Transactions on Engineering Management, 50, (1), 45-63.

Loebbecke, C. (2001). Online Delivered Content: Concept and Potential. E-commerce and V-business. S. J. Barnes and B. Hunt. Oxford, Butterworth-Heinemann.

Loiacono, E. T. (2000). WebQual: A Website Quality Instrument, University of Georgia, Athens.

Loiacono, E. T., Chen, D. O. and Goodhue, D. L. (2002). WebQual Revisited: Predicting the Intent to Reuse a Website. 8th Americas Conference on Information Systems, Barcelona, Spain.

Loiacono, E. T., Watson, R. T. and Goodhue, D. L. (2000). WebQual: A Measure of Web Site Quality, Working Paper 2000-126-0. University of Georgia. 2004.

Lot21 (2001). The Future of Wireless Marketing, http://www.caratinteractive.com/resources/wirless_future.pdf.

Lowe, S. (2003). Read My SMS: Next Generation of Debt is Here., Sydney Morning Herald. 2003.

Magidson, J. (1982) "Some Common Pitfalls in Causal Analysis of Categorical Data", Journal of Marketing Research, 19, (November), 461-471.

Mahajan, V., Muller, E. and Bass, F. M. (1990) "New Product Diffusion Models in Marketing: A Review and Directions for Research", Journal of Marketing, 54, (January), 1-26.

Maignan, I. and Lukas, B. A. (1997) "The Nature and Social Uses of the Internet: A Qualitative Investigation", Journal of Consumer Affairs, 31, (2), 346-371.

Malhotra, N. K. (1991) "Mnemonics in Marketing: A Paedagogical Tool", Journal of the Academy of Marketing Science, 19, (2), 141-149.

Malhotra, N. K., Hall, J., Shaw, M. and Oppenheim, P. (2002), Marketing Research, Prentice Hall, Frenchs Forest.

Manchester, P. (2000). Security Fears are Clouding Heady Growth Forecasts, Financial Times.

Massoud, S. and Gupta, O. (2003) "Consumer Perception and Attitude Toward Mobile Communication", International Journal of Mobile Communications, 1, (4), 390–408.

Mathieson, K. (1991) "Predicting User Intentions: Comparing the Technology Acceptance Model with the Theory of Planned Behavior", Information Systems Research, 2 (3), 173-191.

Matzler, K., Bailom, F., Hinterhuber, H. H., Renzl, B. and Pichler, J. (2004) "The Asymmetric Relationship Between Attribute-Level Performance and Overall Customer Satisfaction: A Reconsideration of the Importance-Performance Analysis", Industrial Marketing Management, 33, (4), 271-277.

Maxwell, J. A. (1996), Qualitative Research Design, SAGE Publications, California.

Maydeu-Olivares, A. (2000) "Review of MPlus", Multivariate Behavioral Research, 35, (4), 501-504.

Mazanec, J. (1982) "Practising the Causal Approach to Consumer Behaviour Model Building: An Example from Tourism Research", Der Markt, 21, (4), 127-133.

Midgley, D. F. and Dowling, G. R. (1978) "Innovativeness: The Concept and Its Measurement", Journal of Consumer Research, 4, (March), 229-242.

Mikami, S., Nakamura, I., Mori, Y., Mori, Y., Korenaga, R., Kenjo, T., Yanagisawa, K. and Sekiya, N. (2001) "Survey Research on Uses of Cellular Phone and PHS in 2000", The Research Bulletin of the Institute of Socio-Information and Communication Studies, The University of Tokyo, 2001, (15), 145-235.

MindMatics (2001) "MindMatics setzt erfolgreiche SMS Kampagne für Coca-Cola und Warner Brothers Movie World um", Press Release, 6 December 2001, (http://www.mindmatics.com/).

Mittal, V., Ross, W. T. and Baldasare, P. M. (1998) "The Asymmetric Impact of Negative and Positive Attribute-Level Performance on Overall Satisfaction and Repurchase Intentions", Journal of Marketing, 62, (January), 33-47.

Monroe, K. B. (1990), Pricing: Making Profitable Decisions, McGraw-Hill, New York.

Montoya-Weiss, M. M., Voss, G. B. and Grewal, D. (2003) "Determinants of Online Channel use and Overall Satisfaction with a Relational, Multichannel Service Provider", Journal of the Academy of Marketing Science, 31, (4), 448-458.

Moon, J.-W. and Kim, Y.-G. (2001) "Extending the TAM for a World-Wide-Web Context", Information & Management, 38, 217-230.

Moore, G. C. and Benbasat, I. (1991) "Development of an Instrument to Measure the Perceptions of Adopting an Information Technology Innovation", Information Systems Research, 2, (3), 192-222.

Morgan, R. M. and Hunt, S. D. (1994) "The Commitment Trust-Theory of Relationship Marketing", Journal of Marketing, 58, (July), 20-38.

Morris, M. and Venkatesh, V. (2000) "Age Differences in Technology Adoption Decisions: Implications for a Changing Work Force", Personnel Psychology, 53, (2), 375-403.

Mplus (2000). Mplus Discussion, Fit Indices for Categorical Outcomes, Various Authors.

Mueller, R. O. (1996), Basic Principles of Structural Equation Modelling: An Introduction to LISREL, Springer, New York.

Mulaik, S. A., James, L. R., Van Alstine, J., Bennett, N., Lind, S. and Stilwell, C. D. (1989) "Evaluation of Goodness-of-Fit Indices for Structural Equation Models", Psychological Bulletin, 105, (3), 430-445.

Müller-Veerse, F. (1999). Mobile Commerce Report. Bonn, Durlacher Research Ltd: 78.

Muthén, L. K. and Muthén, B. (1998), Mplus User's Guide. The Comprehensive Modeling Program for Applied ResearchersLos Angeles.

Myerscough, M. A. (2002). Information Systems Quality Assessment: replicating Kettinger and Lee`s USIF/SERVQUAL Combination. Proceedings of the Eighth Americas Conference on Information Systems (AMCIS).

Nakada, G. (2001). iMode Romps. 2001.

NetValue (2002). The Wireless and Internet Marketing Initiative. Neuilly sur Seine, NetValue. 2001.

Newell, F. (2000), Loyalty.com - Customer Relationship Management in the New Era of Internet Marketing, McGraw-Hill, New York.

Newell, F. and Lemon, K. N. (2001), Wireless Rules - New Marketing Strategies for Customer Relationship Management, McGraw-Hill, New York.

Nomura, A. (2003). Kurumaga keitaikanyushani! KDDI shinsabisunohonmei, Weekly Toyo Keizai. 12/04/2003: 16-17.

Norris, P. (2001), Digital Divide: Civic Engagement, Information Poverty and the Internet in Democratic Societies, Cambridge University Press, New York.

NTT DoCoMo (2000). DoCoMo Report, The Use of Cell Phones/PHS Phones in Everyday Urban Life: A Survey of 1,000 People. Tokyo, NTT DoCoMo, Inc., Public Relations Department.

NTT DoCoMo (2001a). DoCoMo Report Current Trends in Mobile Phone Usage among Adolescents. Tokyo, NTT DoCoMo, Inc., Public Relations Department.

NTT DoCoMo (2001b). DoCoMo Report Mobile Phones Increasingly Popular among the Elderly. Tokyo, NTT DoCoMo, Inc., Public Relations Department.

NTT DoCoMo (2002). DoCoMo Mobile Phone Catalog, English Version Autumn 2002 Edition. Tokyo.

Oliva, T., Oliver, R. L. and MacMillan, I. C. (1992) "A Catastrophe Model for Developing Service Satisfaction Strategies", Journal of Marketing, 56, 83-95.

Oliver, R. L. (1980) "A Cognitive Model of the Antecedents and Consequences of Satisfaction Decisions", Journal of Marketing Research, 17, (November), 460-469.

Oliver, R. L. (1993) "Cognitive, Affective, and Attribute Bases of the Satisfaction Response", Journal of Consumer Research, 20, (December), 418-430.

Oliver, R. L. (1997), Satisfaction: A Behavioral Perspective on the Consumer, McGrawHill, New York.

Oliver, R. L. (1999) "Whence Consumer Loyalty?" Journal of Marketing, 63, 33-44.

Olshavsky, R. W. (1985) "Towards A More Comprehensive Theory of Choice", Advances in Consumer Research, 12, (1), 465-470.

Olson, J. C. and Reynolds, T. J. (1983). Understanding Consumers' Cognitive Structures: Implications for Advertising Strategy. Advertising and Consumer Psychology. L. Percy and A. Woodside. Lexington, MA, Lexington Books.

OMA (2002). Multimedia Messaging Service - Architecture Overview Version 1.1, Open Mobile Alliance. 2002.

Ostrowski, P. L., O'Brien, T. V. and Gordon, G. L. (1993) "Service Quality and Customer Loyalty in the Commercial Airline Industry", Journal of Travel Research, 32, (2), 16-24.

Palmer, J. W. (2002) "Web Site Usability, Design, and Performance Metrics." Information Systems Research, 13, (2), 151.

Parasuraman, A. (1997) "Reflections on Gaining Competitive Advantage Through Customer Value", Journal of the Academy of Marketing Science, 25, (Spring), 154-161.

Parasuraman, A., Berry, L. L. and Zeithaml, V. A. (1991) "Refinement and Reassessment of the SERVQUAL Scale", Journal of Retailing, 67, (4), 420-450.

Parasuraman, A. and Grewal, D. (2000) "The Impact of Technology on the Quality-Value-Loyalty Chain: A Research Agenda", Journal of the Academy of Marketing Science, 28, (1), 168-174.

Parasuraman, A., Zeithaml, V. A. and Berry, L. L. (1985) "A Conceptual Model of Service Quality and Its Implications for Future Research", Journal of Marketing, 49, (Fall), 41-50.

Parasuraman, A., Zeithaml, V. A. and Berry, L. L. (1988) "Servqual: A Multiple-Item Scale for Measuring Consumer Perceptions of Service Quality", Journal of Retailing, 64, (1), 12-40.

Parasuraman, A., Zeithaml, V. A. and Berry, L. L. (1994a) "Alternative Scales for Measuring Service Quality: A Comparative Assessment Based on Psychometric and Diagnostic Criteria", Journal of Retailing, 70, (3), 201-230.

Parasuraman, A., Zeithaml, V. A. and Berry, L. L. (1994b) "Reassessment of Expectations as a Comparison Standard in Measuring Service Quality: Implications for Further Research", Journal of Marketing, 58, (1), 111-124.

Parasuraman, A., Zeithaml, V. A. and Malhotra, A. (2005) "E-S-QUAL A Multiple-Item Scale for Assessing Electronic Service Quality", Journal of Service Research, 7, (3), 213-233.

Pavlou, P. A. (2003) "Consumer Acceptance of Electronic Commerce: Integrating Trust and Risk with the Technology Acceptance Model", International Journal of Electronic Commerce, 7, (3), 101-134.

Paxson, C. (1995) "Increasing Survey Response Rates: Practical Instructions from the Total-Design Method", Cornell Hotel and Restaurant Administration Quarterly, August, 66-74.

Paybox.net (2002). Mobile Payments Delivery Made Simple, White Paper, Version 2.0.

Pearl, J. (2000), Causality - Models, Reasoning, and Inference, Cambridge University Press, Cambridge, USA.

Pedersen, P. E. (2003) "Adoption of Mobile Internet Services: An Exploratory Study of Mobile Commerce Early Adopters", Journal of Organizational Computing and Electronic Commerce, forthcoming.

Pedersen, P. E. and Herbjorn, N. (2003). Usefulness and Self-expressiveness: Extending TAM to Explain the Adoption of a Mobile Parking Service. 16th Bled Electronic Commerce Conference, Bled, Slovenia.

Pedersen, P. E., Leif, M. B. and Thorbjornsen, H. (2002). Understanding Mobile Commerce End-user Adoption: A Triangulation Perspective and Suggestions for an Exploratory Service Evaluation Framework. 35th Hawaii International Conference on System Sciences, Hawaii Big Island, IEEE.

Pedersen, P. E., Nysveen, H. and Thorbjornsen, H. (2002). Adoption of Mobile Services. Model Development and Cross-Service Study., SNF-report no. 31/02. Foundation for Research in Economics and Business Administration, Bergen, Norway.

Peppers, D., Rogers, M. and Dorf, B. (1999) "Is Your Company Ready for One-to-One Marketing?" Harvard Business Review, 77, (1), 151-160.

Petty, R. D. (2000) "Marketing Without Consent: Consumer Choice and Costs, Privacy, and Public Poetry", Journal of Public Policy and Marketing, 19, (Spring), 42-53.

Pitchard, Howard and Havitz (1992) "Loyalty Measurement: A Critical Examination and Theoretical Extension", Leisure Sciences, 14, 155-164.

Pitt, L. F., Watson, R. T. and Kavan, C. B. (1995) "Service Quality: A Measure of Information System Effectiveness", MIS Quarterly, 19, (2), 173-187.

Plouffe, C. R., Hulland, J. S. and Vandenbosch, M. (2001) "Research Report: Richness Versus Parsimony in Modeling Technology Adoption Decisions - Understanding Merchant Adoption of a Smart Card-Based Payment System", Information Systems Research, 12, (2), 208-222.

Plude, D. and Hoyer, W. (1985). Attention and Performance: Identifying and Localizing Age Deficits. Aging and Human Performance. N. Chardness. New York, John Wiley & Sons: 47-99.

Popper, K. R. (1976). Die Logik der Sozialwissenschaften. Der Positivismusstreit in der deutschen Soziologie. T. W. Adorno, R. Dahrendorf, H. Pilot et al. Darmstadt, Luchterhand.

Porter, M. and Millar, V. E. (1985) "How Information Gives You Competitive Advantage", Harvard Business Review, 63, (4), 149-160.

Powell, M. and Vu, D. (2002). Building successful m-commerce businesses.

Puca (2001). Booty Call: How Marketers Can Cross Into Wireless Space. 2001.

Quah, J. T.-S. and Lim, G. L. (2002) "Push selling - Multicast Messages to Wireless Devices Based on the Publish/Subscribe Model", Electronic Commerce Research and Applications, 1, 235-246.

Quios/Engage (2000). The Efficacy of Wireless Advertising: Industry Overview and Case Study, http://www.mobilecommerceworld.com/download/QuiosEngage_WP.pdf.

Raskino, N. (2001). Mobile Coupons Will Reach Right Into Your Pocket, Gartner Group.

Reichheld, F. F. (1993) "Loyalty-based Management", Harvard Business Review, 71, (2), 64-72.

Rodgers, S. and Thorson, E. (2000) "The Interactive Advertising Model: How Users Perceive and Process Online Ads", Journal of Interactive Advertising, 1, (1), http://www.jiad.org/vol1/no1/rodgers/.

Roest, H. and Pieters, R. (1997) "The Nomological Net of Perceived Service Quality", International Journal of Service Industry Management, 8, (4), 336-351.

Rogers, E. M. (1962), Diffusion of Innovations, The Free Press, New York.

Rogers, E. M. (1983), Diffusion of Innovations, Free Press, New York.

Rogers, E. M. (1995), Diffusion of Innovations, The Free Press, New York.

Rubin, H. J. and Rubin, I. S. (1995), Qualitative Interviewing, Sage Publications, Thousand Oaks.

Rubinstein, H. and Griffiths, C. (2001) "Branding Matters More on the Internet", Journal of Brand Management, 8, (6), 394-404.

Ruyter, K. d. and Scholl, N. (1998) "Positioning Qualitative Market Research: Reflections From Theory and Practice", Qualitative Market Research: An International Journal, 1, (1), 7-14.

Sachs, L. (1982), Applied Statistics, Springer-Verlag, New York.

Sadeh, N. (2002), M-Commerce. Technologies, Services and Business Models, Wiley, United States.

Scharl, A. (2000), Evolutionary Web Development, Springer. http://webdev.wu-wien.ac.at/, London.

Scharl, A., Dickinger, A. and Murphy, J. (2005) "Success Factors and Industry Diffusion of Mobile Marketing", Electronic Commerce Research and Applications, 4, (2), 159-173.

Schlenker, B. F., Helm, R. and Tedeschi, J. T. (1973) "The Effects of Personality and Situational Variables of Behavioral Trust", Journal of Personality and Social Psychology, 25, 419-427.

Schleuter, C. and Shaw, M. J. (1997) "A Strategic Framework for Developing Electronic Commerce", IEEE Internet Computing, 1, (6), 20-28.

Schmidt-Belz, B., A., N., Poslad, S. and Zip, A. (2002). Personalized and Location-based Mobile Tourism Services. Mobile HCI, Pisa, Italy.

Schreiber, G. (2000), Schlüsseltechnologie Mobilkommunikation, Deutscher Wirtschaftsdienst, Köln.

Schultz, B. (2001) "The m-commerce fallacy", Network World, 18, (9), 77-82.

Schumacker, R. E. and Lomax, R. G. (1996), A Beginner's Guide to Structural Equation Modeling, Lawrence Erlbaum, Mahwah.

Schuster, T. (2001). Pocket Internet and M-Commerce - (How) Will it Fly?

Seybold, P. (1998), Customer.com, Random House, New York.

Shankar, V., Driscoll, T. and Reibstein, D. (2003) "Rational Exuberance: The Wireless Industry's Killer B", Strategy & Business, 31, (Summer), 68-77.

Sharadanand, U. and Maes, P. (1995). Social Information Filtering: Algorithms for Automating Word-of-Mouth. ACM CHI'95 Conference on Human Factors in Computing Systems, Denver, Colorado.

Shaw, E. (1999) "A Guide to the Qualitative Research Process: Evidence from a Small Firm Study", Qualitative Market Research: An International Journal, 2, (2), 59-70.

Sheth, J. N., Newman, B. I. and Gross, B. L. (1991), Consumption Values and Market Choices: Theory and Applications, Southwestern Publishing, Cincinnati, OH.

Sirohi, N., McLaughlin, E. W. and Wittink, D. R. (1998) "A Model of Consumer Perceptions and Store Loyalty Intentions for a Supermarket Retailer", Journal of Retailing, 74, (2), 233-245.

Spreng, R. A. and Mackoy, R. D. (1996) "An Empirical Examination of a Model of Perceived Service Quality and Satisfaction", Journal of Retailing, 72, (2), 201-214.

Steenkamp, J.-B. E. M. and Baumgartner, H. (1998) "Assessing Measurement Invariance in Cross-National Consumer Research", Journal of Consumer Research, 25, 78-90.

Steenkamp, J.-B. E. M. and Baumgartner, H. (2000) "On the Use of Structural Equation Models for Marketing Modeling", International Journal of Research in Marketing, 17, 195-202.

Steenkamp, J.-B. E. M. and Baumgartner, H. (2001) "Response Styles in Marketing Research: A Cross National Investigation", Journal of Marketing Research, 38, (May), 143-156.

Stein, L. D. (1998), Web Security, Addison-Wesley, Reading, MA.

Stratil, A. and Weissenburger, E.-M. (2000), Telekommunikationsgesetz, Manz, Vienna.

Straub, D., Keil, M. and Brenner, W. (1997) "Testing the Technology Acceptance Model Across Cultures: A Three Country Study", Information & Management, 31, (1), 1-11.

Sudman, S. and Bradburn, N. M. (1982), Asking Questions: A Practical Guide to Questionnaire Design, Jossey-Bass, San Francisco.

Symonds, M. (1999). Business and the Internet: Survey, Economist. 26: 1-44.

Tapscott, D. (1995), The Digital Economy, McGraw-Hill, New York.

Taylor, S. and Todd, P. (1995a) "Assessing IT Usage: The Role of Prior Experience", MIS Quarterly, 4, (December), 561-570.

Taylor, S. and Todd, P. A. (1995b) "Understanding Information Technology Usage: A Test of Competing Models." Information Systems Research, 6, 144-176.

Tema-Lyn, L. (1999) "Five Ways to Get More Out of Qualitative Research", Marketing News, 33, (12), 38.

Teo, T., S. H., Lim, V. K. G. and Lai, R. Y. C. (1999) "Intrinsic and Extrinsic Motivation in Internet Usage", International Journal of Management Science, 27, 25-37.

Thaler, R. (1985) "Mental Accounting and Consumer Choice", Marketing Science, 4, (Summer), 199-214.

Thiele, C. and Liess, A. (2000). Social Impact Study. Vienna, Fessel-GfK Market Research.

Thompson, R. L., Higgins, C. A. and Howell, J. M. (1991) "Personal Computing: Toward a Conceptual Model of Utilization", MIS Quarterly, 15 (1), 125-143.

Ticoll, D., Lowy, A. and Kalakota, R. (1998). Joined at the Bit - the Emergence of the e-Business Community. Blueprint to the Digital Economy. D. Tapscott, A. Lowy and D. Ticoll. New York, McGraw-Hill.

Tornatzky, L. G. and Klein, K. J. (1982) "Innovation Characteristics and Innovation Adoption - Implementation: A Meta-Analysis of Findings", IEEE Transactions on Engineering Management, 29, 28-45.

Triand's, H. C. (1977), Interpersonal Behavior, Brooke/Cole, Monterey, CA.

Triandis, H. C. (1980). Values, Attitudes and Interpersonal Behavior. Nebraska Symposium on Motivation, Beliefs, Attitudes and Values, Lincoln, NE, University of Nebraska Press.

Tsikriktsis, N. (2002) "Does Culture Influence Web Site Quality Expectations? An Empirical Study", Journal of Service Research, 5, (2), 101-112.

Vallerand, R. J. (1997). Toward a Hierarchical Model of Intrinsic and Extrinsic Motivation. Advances in Experimental Social Psychology. M. Zanna. New York, Academic Press. 29: 271-360.

Van den Bulte, C. and Lilien, G. L. (2001) "Medical Innovation Revisited: Social Contagion versus Marketing Effort", American Journal of Sociology, 106, (5), 1409-1435.

Van der Heijden, H. (2003) "Factors Influencing the Usage of Websites: The Case of a Generic Portal in The Netherlands", Information & Management, 40, 541-549.

Van der Heijden, H. (2004) "User Acceptance of Hedonic Information Systems", MIS Quarterly, 28, (4), 695-704.

Varshney, U. (2003) "Location Management for Mobile Commerce Applications in Wireless Internet Environment", ACM Transactions on Internet Technology, 3, (3), 236-255.

Varshney, U. and Vetter, R. (2001). A Framework for the Emerging Mobile Commerce Applications. 34th Hawaii International Conference on System Sciences, Hawaii, IEEE Computer Science Press, Los Alamitos.

Venkatesh, V. (1999) "Creation of Favorable User Perceptions: Exploring the Role of Intrinsic Motivation", MIS Quarterly, 23, (2), 239-260.

Venkatesh, V. (2000) "Determinants of Perceived Ease of Use: Integrating Control, Intrinsic Motivation, and Emotion into the Technology Acceptance Model", Information Systems Research, 11, (4), 342-356.

Venkatesh, V. and Davis, F. D. (2000) "A Theoretical Extension of the Technology Acceptance Model: Four Longitudinal Field Studies", Management Science, 46, (2), 186-204.

Venkatesh, V. and Morris, M. (2000) "Why Don't Men Ever Stop to Ask for Directions? Gender, Social Influence and their Role in Technology Acceptance and Usage Behavior", MIS Quarterly, 24, (1), 115-139.

Venkatesh, V., Morris, M. G., Davis, G. B. and Davis, F. D. (2003) "User Acceptance of Information Technology: Toward a Unified View", MIS Quarterly, 27, (3), 425-478.

Venkatesh, V. and Speier, C. (1999) "Computer Technology Training in the Workplace: A Longitudinal Investigation of the Effect of the Mood", Organizational Behavior and Human Decision Processes, 79, (1), 1-28.

von Hippel, E. (1986) "Lead Users: A Source of Novel Product Concepts", Management Science, 32, 791-805.

von Hippel, E. (1999). Toolkits for User Innovation, MIT Sloan School of Management Working Paper No. 4058.

Voss, K. E., Spangenberg, E. R. and Grohmann, B. (2003) "Measuring the Hedonic and Utilitarian Dimensions of Consumer Attitude", Journal of Marketing Research, XL, (August 2003), 310-320.

Wales, E. (2003) "Industry Sinks Teeth into Spam", Network Security, 9, 15-17.

Wang, Y.-S. and Tang, T.-I. (2003) "Assessing Customer Perceptions of Website Service Quality in Digital Marketing Environments", Journal of End User Computing, 15, (3), 14-31.

Watson, R. T. (2004) "I Am My Own Database", Harvard Business Review, November, 18-19.

Watson, R. T., Berthon, P., Pitt, L. F. and Zinkhan, G. M. (2000), Electronic Commerce: The Strategic Perspective, Dryden, Fort Woth, TX.

Watson, R. T., Pitt, L. F., Berthon, P. and Zinkhan, G. M. (2002) "U-Commerce: Expanding the Universe of Marketing", Journal of the Academy of Marketing Science, 30, (4), 333-347.

Watson, R. T., Pitt, L. F. and Kavan, C. B. (1998) "Measuring Information Systems Service Quality: Lessons From Two Longitudinal Case Studies", MIS Quarterly, 22, (1), 61-79.

Webster, J. and Trevino, L. K. (1995) "Rational and Social Theories as Complementary Explanations of Communication Media Choices: Two Policy-capturing Studies", Academy of Management Journal, 38, 1544-1572.

Weinhold-Stünzi, H. (1994) "Die Kunst der Markt- und Meinungsforschungen oder: Betrachtungen zum Vergleich repräsentativer Befragungen mit intersubjektiven Expertenerhebungen", Thesis: Marktforschung, 99-104.

Wells, J. D., Sarker, S., Urbaczewski, A. and Sarker, S. (2002). Studying Customer Evaluations of Electronic Commerce Applications: A Review and Adaptation of the Task-Technology Fit Perspective. Hawaii International Conference on System Sciences, Hawaii, IEEE The Computer Society.

Whyte, W. H. J. (1983) "The Strength of Weak Conversational Ties in the Flow of Information and Influence", Social Networks, 5, 245-267.

Wolfinbarger, M. and Gilly, M. C. (2003) "eTailQ: Dimensionalizing, Measuring and Predicting eTail Quality", Journal of Retailing, 79, 183-198.

Woodruff, R. B. (1997) "Customer Value: The Next Source of Competitive Advantage", Journal of the Academy of Marketing Science, 25, (2), 139-135.

Woodruff, R. B., Cadotte, E. R. and Jenkins, R. L. (1983) "Modeling Consumer Satisfaction Processes Using Experience-Based Norms", Journal of Marketing Research, 20, 296-304.

Woodruff, R. B. and Gardial, S. F. (1996), Know your Customer: New Approaches to Understanding Customer Value and Satisfaction, Blackwell Publishers, Cambridge, MA.

Wu, J.-H. and Wang, S.-C. (2005) "What Drives Mobile Commerce? An Empirical Evaluation of the Revised Technology Acceptance Model", Information & Management, 42, (5), 719.

WWRF (2000). The Book of Visions 2000 - Visions of the Wireless World, IST-WSI Project Report version 1.0, IST-WSI/WWRF.

Xu, Y. (1999). Development of Transport Telematics in Europe. International Workshop on Geographic Information Systems for Transportation (GIS-T) and Intelligent Transportation Systems (ITS), Hong Kong.

Yoo, B. and Donthu, N. (2001) "Developing a Scale to Measure the Perceived Quality of Internet Shopping Sites (SITEQUAL)", Quarterly Journal of Electronic Commerce, 2, (1), 31-47.

Young, S. and Feigen, B. (1975) "Using the Benefit Chain for Improved Strategy Formulation", Journal of Marketing, 39, (July), 72-74.

Yu, C. Y. (2002). Evaluating Cutoff Criteria of Model Fit Indices for Latent Variable Models with Binary and Continuous Outcomes. Los Angeles, University of California.

Yunos, H. M., Gao, J. Z. and Shim, S. (2003) "Wireless Advertising's Challenges and Opportunities", IEEE Computer, 36, (5), 30-37.

Zeithaml, V. A. (1988) "Consumer Perceptions of Price, Quality and Value: A Means-End Model and Synthesis of Evidence", Journal of Marketing, 52, (3), 2-22.

Zeithaml, V. A., Berry, L. L. and Parasuraman, A. (1996) "The Behavioral Consequences of Service Quality", Journal of Marketing, 60, (April), 31-46.

Zeithaml, V. A. and Bitner, M. J. (1996), Services Marketing, The McGraw-Hill Companies Inc., New York.

Zeithaml, V. A., Parasuraman, A. and Malhotra, A. (2000). A Conceptual Framework for Understanding e-Service Quality: Implications for Future Research and Managerial Practice, Working Paper No. 00-115, Marketing Science Institute, Cambridge, MA.

Zhang, Y. and Im, I. (2002). Recommender Systems: A Framework and Research Issues. Americas Conference on Information Systems.

Zirgus, I. and B.K., B. (1998) "A Theory of Task Technology Fit and Group Support System Effectiveness", MIS Quarterly, 22, (3), 313-334.

Zmud, R. W. and Apple, L. E. (1992) "Measuring Technology Incorporation/Infusion", Journal of Product Innovation Management, 9, (June), 148-155.

Zobel, J. (2001), Mobile Business und M-Commerce, Die Märkte der Zukunft erobern, Carl Hanser Verlag, München, Wien.

12 APPENDIX - QUESTIONNAIRE

Bei welchem Betreiber ist Ihr Handy gemeldet?
T-Mobile (0676)
A1 (0664)
One (0699)
Tele Ring (0650)
Drei (0660)
Yess (0699)

Bitte kreuzen Sie alle mobilen Dienste an, die Sie

	...bereits genutzt/bekommen haben	... gerne nützen/bekommen würden
Informationsdienste (Wetter, Sport,..)	O	O
Finanzdienstleistungen (Bank, Börse)	O	O
Ticket-Kauf (Kino, Konzert,...)	O	O
Werbung	O	O
Standortbezogene Dienste (Location Based Services, nächster Bankomat, Restaurant...)	O	O
Unterhaltung (Spiele, Klingeltöne, Musik,..)	O	O
Fahrschein	O	O
Shopping	O	O
Lottoschein	O	O
m-parking	O	O
Sonstige	O	O

Wie oft haben Sie mobile Dienste genutzt?
Öfter als 1 mal pro Woche
1 Mal pro Woche
1 Mal pro Monat
1 Mal pro Jahr
Noch Nie

Seit wann nutzen Sie mobile Dienste?
>3 Jahre
2 Jahre
1 Jahr
0,5 Jahre
Gar nicht

Haben Sie schon einmal M-Parking genutzt?
Ja
Nein (Skip logic, continue with questions on demographics)

Wie sind Sie auf m-parking aufmerksam geworden?
Fernsehen
Radio
Freunde
E-Mail
Vorträge
Internetrecherche
Zeitungen
Fachzeitschriften
Sonstiges

Wie haben Sie sich für m-parking angemeldet?
SMS
Internet
Call Center
WAP

Was war der Hauptgrund sich tatsächlich für m-parking anzumelden?
Gezielte Werbeaktion
Gewinnspiel
Freunde
Neugier
Ärger über Parkscheine
Sonstiges

In welchem Ort nutzen sie m-parking?

Bregenz	Gleisdorf	Mödling
St. Pölten	Stockerau	Wels
Wien	Bludenz	Kitzbühel
Krems	Tulln	Amstetten

Bitte geben Sie an, wie sehr Sie folgenden Aussagen zu m-parking zustimmen?

	Sehr	Ziemlich	Weniger	Eher nicht
Der Dienst hat Premium-Qualität.	o	o	o	o
Der Dienst erfüllt höchste Ansprüche.	o	o	o	o
Der Dienstleistungsstandard ist gut.	o	o	o	o

	Sehr gut	Gut	Eher nicht gut	Schlecht
Bitte geben Sie an, welchem Gefühl Ihre Qualitätseinschätzung von M-Parking entspricht.	o	o	o	o
Bitte geben Sie an, wie zufrieden Sie mit dem Dienst sind.	o	o	o	o

Wie hoch bzw. niedrig schätzen Sie den Nutzen bzw. den Aufwand in Verbindung mit der Nutzung von m-parking ein?

	Hoch			Niedrig
Insgesamt ist der Nutzen des Dienstes für mich:	o	o	o	o
Angesichts dessen, wieviel ich für den Dienst bezahle, ist der Nutzen:	o	o	o	o

Im Vergleich dazu, wie schwierig die Bedienung ist, ist der Nutzen:	O	O	O	O
Der Betrag, den ich für diesen Dienst bezahle ist:	O	O	O	O
Der Zeitaufwand, um den mobilen Dienst zu nutzen, ist:	O	O	O	O
Die Mühe, um den mobilen Dienst zu nutzen, ist:	O	O	O	O

Bitte geben Sie an, wie sehr Sie folgenden Aussagen zu m-parking zustimmen:

	Stimme voll zu			Stimme nicht zu
Es war einfach zu lernen, wie der Dienst funktioniert.	O	O	O	O
Es ist einfach zu schaffen, dass der Dienst tut, was ich will.	O	O	O	O
Meine Interaktion mit dem Dienst ist einfach und verständlich.	O	O	O	O
Der Dienst ist einfach zu bedienen.	O	O	O	O
Der Dienst ermöglicht es mir, die Transaktion rasch auszuführen.	O	O	O	O
Der Dienst hilft mir Zeit zu sparen.	O	O	O	O
Die Nutzung des Dienstes ist bequem für mich.	O	O	O	O
Meine Kollegen und Freunde meinten, dass ich diesen mobilen Dienst nutzen soll.	O	O	O	O
Leute, die ich kenne, dachten, es ist eine gute Idee diesen Dienst zu nutzen.	O	O	O	O
Leute, die ich kenne, beeinflussten mich diesen Dienst auszuprobieren.	O	O	O	O
Es ist ein guter Zeitvertreib, diesen Dienst zu nutzen.	O	O	O	O
Es ist unterhaltsam, diesen Dienst zu nutzen.	O	O	O	O
Es macht Spaß, diesen Dienst zu nutzen.	O	O	O	O
Der Dienst steht immer zur Verfügung.	O	O	O	O
Die Übertragung der Nachrichten bricht nicht ab.	O	O	O	O
Ich habe das Gefühl, meine Privatsphäre wird vom Serviceanbieter geschützt.	O	O	O	O
Ich fühle mich bei meinen Transaktionen mit dem Serviceanbieter sicher.	O	O	O	O
Ich plane, diesen Dienst wieder zu nutzen	O	O	O	O
Ich werde weitere mobile Dienste dieses Anbieters nutzen.	O	O	O	O
Ich würde zu einem anderen Betreiber wechseln, der den Dienst auch anbietet.	O	O	O	O
Ich würde zu einem anderen Betreiber wechseln, der den Dienst günstiger anbietet.	O	O	O	O
Im Freundeskreis bin ich gewöhnlich unter den ersten, die technische Produkte ausprobieren.	O	O	O	O

Ich mag es, mit neuen technischen Produkten zu experimentieren.	O	O	O	O
Ich werde anderen gegenüber positiv über diesen Dienst sprechen.	O	O	O	O
Ich werde den Dienst empfehlen, wenn jemand meinen Rat sucht.	O	O	O	O
Ich werde Freunde ermuntern, den Dienst zu nutzen.	O	O	O	O
Ich bin auf Grund von vergangener Nutzung mit dem Dienst vertraut.	O	O	O	O
Ich würde mehr zahlen, wenn ich unter Zeitdruck bin.	O	O	O	O
Ich habe bei der Ersten Verwendung des Dienstes eine gute Erfahrung gemacht.	O	O	O	O
Der Dienst steht jederzeit zur Verfügung.	O	O	O	O
Der Dienst steht überall zur Verfügung.	O	O	O	O

Bitte machen Sie noch einige Angaben zur Person.

Geschlecht:	Männlich	O Weiblich	
Wie alt sind Sie?____			
Was ist Ihre höchste abgeschlossene Ausbildung?	Pflichtschule Fachschule/Lehre	Matura Hochschule	
Was ist Ihr Beruf?	Pensionist/in In Ausbildung Selbstständig Arbeiter/in Angestellte/r	Hausfrau/mann Landwirt/in Derzeit ohne Beschäftigung Student/in Beamte/r	
Wie viele Einwohner hat der Ort, aus dem Sie kommen?	Bis 1.000 1.001-5.000 5.001 - 10.000 10.001 - 50.000	50.000 - 100.000 100.001 - 500.000 500.001 - 1 Mio. Mehr als 1 Mio.	
Was ist Ihre Postleitzahl? _____			
Wie hoch ist Ihr monatliches Netto- Einkommen?	0-1000 Euro 1001-2000 Euro	2001-3000 Euro Darüber	

Vielen Dank für Ihre Unterstützung bei dieser Studie!

Forschungsergebnisse der Wirtschaftsuniversität Wien

Herausgeber: Wirtschaftsuniversität Wien –
vertreten durch a.o. Univ. Prof. Dr. Barbara Sporn

Band 1 Stefan Felder: Frequenzallokation in der Telekommunikation. Ökonomische Analyse der Vergabe von Frequenzen unter besonderer Berücksichtigung der UMTS-Auktionen. 2004.

Band 2 Thomas Haller: Marketing im liberalisierten Strommarkt. Kommunikation und Produktplanung im Privatkundenmarkt. 2005.

Band 3 Alexander Stremitzer: Agency Theory: Methodology, Analysis. A Structured Approach to Writing Contracts. 2005.

Band 4 Günther Sedlacek: Analyse der Studiendauer und des Studienabbruch-Risikos. Unter Verwendung der statistischen Methoden der Ereignisanalyse. 2004.

Band 5 Monika Knassmüller: Unternehmensleitbilder im Vergleich. Sinn- und Bedeutungsrahmen deutschsprachiger Unternehmensleitbilder – Versuch einer empirischen (Re-)Konstruktion. 2005.

Band 6 Matthias Fink: Erfolgsfaktor Selbstverpflichtung bei vertrauensbasierten Kooperationen. Mit einem empirischen Befund. 2005.

Band 7 Michael Gerhard Kraft: Ökonomie zwischen Wissenschaft und Ethik. Eine dogmenhistorische Untersuchung von Léon M.E. Walras bis Milton Friedman. 2005.

Band 8 Ingrid Zechmeister: Mental Health Care Financing in the Process of Change. Challenges and Approaches for Austria. 2005.

Band 9 Sarah Meisenberger: Strukturierte Organisationen und Wissen. 2005.

Band 10 Anne-Katrin Neyer: Multinational teams in the European Commission and the European Parliament. 2005.

Band 11 Birgit Trukeschitz: Im Dienst Sozialer Dienste. Ökonomische Analyse der Beschäftigung in sozialen Dienstleistungseinrichtungen des Nonprofit Sektors. 2006

Band 12 Marcus Kölling: Interkulturelles Wissensmanagement. Deutschland Ost und West. 2006.

Band 13 Ulrich Berger: The Economics of Two-way Interconnection. 2006.

Band 14 Susanne Guth: Interoperability of DRM Systems. Exchanging and Processing XML-based Rights Expressions. 2006.

Band 15 Bernhard Klement: Ökonomische Kriterien und Anreizmechanismen für eine effiziente Förderung von industrieller Forschung und Innovation. Mit einer empirischen Quantifizierung der Hebeleffekte von F&E-Förderinstrumenten in Österreich. 2006.

Band 16 Markus Imgrund: Wege aus der Insolvenz. Eine Analyse der Fortführung und Sanierung insolventer Klein- und Mittelbetriebe unter besonderer Berücksichtigung des Konfigurationsansatzes. 2007.

Band 17 Nicolas Knotzer: Product Recommendations in E-Commerce Retailing Applications. 2007.

Band 18 Astrid Dickinger: Perceived Quality of Mobile Services. A Segment-Specific Analysis. 2007.

www.peterlang.de

Jens Ulrich Hanisch

Rounding of Income Data
An Empirical Analysis of the Quality of Income Data with Respect to Rounded Values and Income Brackets with Data from the European Community Household Panel

Frankfurt am Main, Berlin, Bern, Bruxelles, New York, Oxford, Wien, 2007.
XXIV, 246 pp., num. tab. and graphs
Schriften zur empirischen Wirtschaftsforschung.
Herausgegeben von Peter M. Schulze. Bd. 9
ISBN 978-3-631-55687-0 · pb. € 45.50*
US-ISBN 0-8204-8721-X

Income questions are frequently answered with rounded values or income brackets. This has an impact on the quality of data, which is demonstrated for the European Community Household Panel (ECHP) and the German Socio-Economic Panel (SOEP). A matching of register and interview data for the Finnish sub-sample of the ECHP allows an analysis of the measurement error caused by rounding with regard to cross-sectional statistics and the mobility of incomes. The emphasis is on income quantiles, poverty measures and income mobility. The finding is that most income values are rounded after one or two significant digits, and the accuracy improves only slightly after the initial wave. The results are that rounding behaviour can change across panel waves, and can also be different across countries and types of income. Characteristics like gender, job type and mode of interview were significantly correlated with rounding behaviour.

Contents: Surveys contain rounded income values · Problems when using rounded values · Association of rounding behaviour with other factors in the ECHP and SOEP data

Frankfurt am Main · Berlin · Bern · Bruxelles · New York · Oxford · Wien
Distribution: Verlag Peter Lang AG
Moosstr. 1, CH-2542 Pieterlen
Telefax 00 41 (0) 32 / 376 17 27

*The €-price includes German tax rate
Prices are subject to change without notice
Homepage http://www.peterlang.de